BRITAIN'S WEATHER

BRITAIN'S WEATHER

Its Workings, Lore and Forecasting

DAVID BOWEN

DAVID & CHARLES : NEWTON ABBOT

1969

7153 4717 9

This book is set in 11 on 13 pt Times Roman

Printed in Great Britain by
Clarke Doble & Brendon Limited Plymouth
for David & Charles (Publishers) Limited
South Devon House Newton Abbot Devon

CONTENTS

LIST OF ILLUSTRATIONS

PLATES

Unless otherwise stated, the photographs are from the David Bowen Cloud & Weather Library.

IN TEXT

1

THE CLOUDS

Men judge by the complexion of the sky
The state and inclination of the day.

THIS was Shakespeare's observation (Richard II, iii 2) of the use of a craft many thousands of years old—the reading of cloud signs in the sky. To the patient observer clouds are not merely a part of the present weather scene, they are also indications of what may happen in the near future; the well known meteorologist Sir Napier Shaw referred to them as actors which, although they never repeat their performances exactly, do possess recognisable traits.

Local forecasts can be made by noting the cloud types, whether they are increasing or decreasing in size, their speed of movement, the time of day, the season, and the general features of the area. In olden times shepherds, mariners and others did this with uncanny skill, though without the help of science. Today's sky watchers favour an approach that is as far as possible scientific as well as practical. Any short course in meteorology or radio talk on the subject will have told them that before clouds can be produced the air must contain a certain proportion of moisture and that this, in fact, is nearly always the case; the amount varies from day to day and place to place but it is only over very arid deserts, and not always then, that air is ever completely dry.

The second essential for the production of cloud is that the air should be chilled. This happens very easily, for example, when winds blow over land or sea surfaces that are colder than they are, or when air expands, as it inevitably does, after being forced to rise; and air rises when it is heated by ground or water surfaces or when it is blown against an obstruction like the side of a hill.

Now what is particularly interesting is that the colder any mass of air happens to be, the less moisture it can contain in the invisible form. Cloud will appear when it is cooled to below a certain critical temperature known as its dew point, and this will vary according to the amount of moisture present. But why is it that some clouds produce rainfall and some do not?

The answer is that a cloud must grow at least moderately large before it can produce rain, snow, or hail. To the observer at ground level, who cannot tell the thickness of a cloud, it may be sufficient to regard it as capable of producing rainfall if its base is dark and covers at least half the sky between one horizon and another. But remember that on a hazy day the general atmospheric obscurity will tend to make cloud bases appear darker and more threatening than they really are.

A seemingly fantastic amount of water is suspended in the air currents within any cloud. Sir Graham Sutton, a recent director of the Meteorological Office, calculated it to be anything from 100 to 1,000 tons for a small cloud, and for a large shower cloud something in the region of 100,000 tons.[1] Just how much water must be contained in the clouds of a general storm area is difficult to imagine as well as to calculate.

For rain to fall to the ground the very minute cloud drops must be converted to much larger ones, and an average raindrop in fact represents the aggregate, according to Sutton, of about 1,000,000 cloud drops. How this happens is still not perfectly understood, but there are two theories today that are widely accepted.

The first one, put forward by the Swedish meteorologist T. Bergeron, takes into account the fact that a large proportion of rainfall begins as snow. Near the top of a rain-producing cloud formation temperatures are likely to be below −10° C and many of the cloud droplets are 'supercooled', that is, still liquid despite the sub-zero temperature. But there will also be some ice crystals present at this temperature. These are likely to grow by chain reaction at the expense of the water drops, which will freeze on contact with them, become snowflakes, and, when heavy enough, fall by gravity to lower levels. Finally, unless temperatures here are below freezing, they will melt to become raindrops.

The process has been demonstrated under laboratory conditions. In addition, modern cloud-radar has proved that many of the rain clouds of temperate latitudes have snowflakes in their upper layers.

However, it is known that showers can fall from clouds that are too warm, even in their topmost layers, to contain supercooled drops; this is often the case with the very large shower cloud formations found in equatorial regions. Here the raindrops are probably formed from the original small cloud drops by coalescence, and the violent winds inside these cloud formations are thought to be capable of aiding this process.

Basically, there are really only two types of cloud: the heaped cloud of the *cumulus* family and the layer cloud of the *stratus* family. These major types are conveniently subdivided according to the height at which they occur: the highest (base 15,000 ft and above) called *cirrus*, or prefixed *cirro-*; the clouds of medium height (base between 8,000 and 15,000 ft) prefixed *alto-*; and the low clouds (base lower than 8,000 ft), in which group are included clouds which despite their low bases reach up high into the atmosphere.

Cumulus is the most common cloud formation, and it is usually easy to recognise, having a flat base that is normally about 2,000–3,000 ft above the ground, and a rounded top; in colour it is a brilliant white when the sun is shining on it, but it appears as a light, medium or dark shade of grey (according to its thickness) when directly overhead, casting shadows on the ground. It has another distinctive quality: whatever its size, gaps normally appear between the individual clouds, through which can be seen either blue sky or clouds at a higher level.

Within these broad limits, however, *cumulus* formations differ widely. The smaller clouds occur during periods of fine weather; the larger ones can produce showers. The largest of all may reach a height of 15,000 ft or more and will appear mountainous and threatening. Hence the saying:

> When mountains and cliffs in the sky appear
> Some sudden and violent showers are near.

The largest clouds in the *cumulus* family, called *cumulonimbus*, are some 20,000 ft or more in vertical thickness and give heavy showers, sometimes with hail, snow flurries, and lightning; they can often be recognised from a distance by characteristic anvil-shaped tops that spread out laterally across the sky. The ancient Greeks referred to these as 'white tempests'.

Individual *cumulus* or *cumulonimbus* shower clouds, when they pass, usually allow good bright intervals before the sky darkens again with the approach of another shower, but there are days when a number of these clouds are blown together by strong winds, and the breaks in the sky between them are only very small. The gaps between the showers or overcast periods are then only momentary. At such times little or no cloud top is visible from the ground, and the clouds may be difficult to recognise until—probably towards evening—the wind strength moderates and allows the formations to separate. Often, in inland regions—and this is the case with all *cumulus* formations—the clouds will rapidly decrease in size during the late afternoon and may disappear altogether by nightfall.

Large *cumulus* and *cumulonimbus* clouds may occur on warm and potentially rather unsettled, thundery days during the summer, and will mass together before the onset of any period of great thunderstorm activity. Visibility will be generally poor before and during the thundery rainfall. But the formations are also common on days during any season when winds are fresh and blowing from the Atlantic, generally from a point between west and north-west, and visibility will then be poor during showers but good between them.

On such days western districts will usually have the greatest rainfall, which will be particularly heavy on exposed hill slopes where the inblowing winds are being forced to rise to higher levels than they would do otherwise. Inland, and away from the prevailing wind, the amount of rainfall will be smaller and the showers generally less frequent. Here showery days may begin with clear blue skies, though a quick increase in the strength of the wind and the early appearance of swiftly moving cloud fragments will indicate that the weather is not to be trusted. The clearing of the clouds in inland regions towards evening is not in itself a sign that the weather is becoming more settled.

Stratocumulus, which, as its name implies, is a mixture of *stratus* and *cumulus*, is more of a layer than a heaped cloud formation. When one is confronted early in the day with a layer of this cloud it is difficult to make an immediate forecast without knowing the weather trend over the previous day or two. But rain may be indicated within a few hours if the rolls or globules of *stratocumulus* join together and become progressively darker. More often than not this does not happen, and the formation breaks up now and again during the day to allow sunny periods; the formation, however, tends to be somewhat more persistent in coastal areas, where winds are blowing from the sea, than elsewhere.

Stratus cloud differs from *stratocumulus* in that it has poor definition, appearing rather as a fog above the ground. If it descends to cover the ground, then it is officially classed as fog, but it is always described as *stratus* so long as it remains clear of the lowest ground in the vicinity of the observer.

After a fine warm day *stratus* may form during the evening, but it will probably disappear soon after sunrise on the following morning. In winter it is generally associated with foggy or hazy weather and may produce light rain or drizzle. If *stratus* forms during the morning over high ground, and hangs about, this is generally a sign of a rapid deterioration in the weather at any season. An exception to this rule is when fog that has formed overnight lifts to become *stratus* cloud for a short time before disappearing, but this generally occurs only between late spring and early autumn.

In striking contrast are the clouds in the high *cirrus* group, which frequently indicate bad weather. A fine cottony *cirrus* may occur during periods of settled weather, but if it becomes thicker, tufted, or shows pronounced waves or bands, or again if it gives way to a milky white sheet *(cirrostratus)* at a similar height, then rain accompanied by strong wind can be expected within 5–6 hours. This development indicates the approach of a 'warm front' associated with a low-pressure weather system. (Weather systems that affect the British Isles are dealt with in Chapter 4.) The deterioration in the weather will be gradual, spread over a period of several hours at least. However, the rain is likely to be continuous once it sets in, which accounts for the saying, 'long notice, long past'. The converse of this saying, 'short notice, soon past', applies to the showers associated with *cumulus* and *cumulonimbus* formations.

Sometimes the gradual deterioration in the weather may remain unnoticed for some time, if the high tufted or wavy *cirrus*, or the *cirrostratus* bands or sheets, are obscured by lower cloud formations, though they can usually be detected if there are any convenient gaps in these lower clouds. At night the task is more difficult, but the presence of a high *cirrostratus* sheet can be assumed if there is a halo, or if the moon appears watery or blurred but with its outline still visible. Rain can then be forecast with some certainty.

There is confirmation of the approaching rain if the winds freshen from a southerly direction and if the *cirrostratus* is succeeded by an *altostratus* sheet, which is, as the name indicates, a cloud of medium height. In appearance it is similar to the *cirrostratus*, only darker. At first the moon or the sun shining through it will appear as though seen through ground glass, their outlines not visible; but within 1–2 hours the cloud

Page 17: *(above)* Small *cumulus* clouds of fair weather; *(below)* roll of *stratocumulus*. This normally occurs during fair weather, allowing sunny periods of varying length.

Page 18: A potential shower cloud. This *cumulus* cloud, with its protruding 'cauliflower' top, is growing vertically.

will thicken and become opaque. Rain is not likely to reach the ground in any quantity until the *altostratus* gives way to heavy-looking *nimbostratus*, whose base may be no more than 1,500 ft above the ground, and will appear a very dark grey. If the temperature is below 37° F then sleet or snow, and not rain, is likely to fall from this cloud, which will assume a characteristic zinc-grey appearance.

Altocumulus, which occurs at approximately the same height as *altostratus*, and which is a light or medium grey in colour according to the relative position of the sun, is less menacing and does not always indicate a deterioration in the weather. The clouds in this formation may occur in waves, like high *stratocumulus*, or lens-shaped, or like a banner. The last two types are often associated with gradually improving weather over hilly or mountainous country, though the elongated shape is caused by very strong winds that are blowing in the area where the cloud is being formed.

So-called 'mackerel' skies may be either *altocumulus* or, rather more rarely, *cirrocumulus*, which appears at high-cloud level and can be recognised by its minute cloud globules that are pure white in colour. The weather trend indicated by this formation is sometimes a matter of dispute, but the saying:

> Mackerel sky, mackerel sky,
> Never long wet and never long dry

is generally true and easy to remember.

2

THE ELEMENTS

IN the meteorological world deposits of water, whether in liquid or in solid form, are collectively known as 'precipitation', but this rather ugly word is seldom used in the daily forecasts. Instead, the forecaster refers to the individual elements as and when they are expected, although from time to time one may hear him mention two elements together—for example, 'rain or snow' or 'rain or drizzle'—if there is any doubt as to which of the respective elements is expected. If we talk—as we often tend to do—of the amount of 'rainfall' of any region when describing its general climate, we mean anything that can be measured in a rain gauge, and this includes not only the water that falls from the sky but also other types of precipitation such as deposits of dew, and melted hoar frost (frozen dew), snow and hail. The measurements are made in inches or millimetres and are normally quoted in terms of averages for each district over a period of at least thirty-five years.

Various scientific attempts have been made to measure raindrops, which differ greatly in size. The largest of them may be as much as 5½ mm in diameter (about 0·2 in), but if they try to grow any larger than this the air resistance will break them up as they fall. The raindrops that increase in size with the greatest rapidity are generally associated with the very fierce air currents inside *cumulonimbus* clouds. Now and again these currents may suddenly be halted at a particular point, leading to a larger than normal proportion of the cloud's water content falling rapidly to the ground. These often disastrous events are known familiarly as 'cloudbursts'.

At the other extreme, if raindrops are very small and numerous, the precipitation is then known as drizzle. Countrymen sometimes refer to it as 'mizzly rain'. The distinction between rain and drizzle is not easy to express precisely, though it has been suggested, but not internationally agreed, that drops of only 0·5 mm in diameter should be known as drizzle. In practice it is known that drops of drizzle are normally considerably smaller than this, and however wet drizzle may appear to those who walk through it the amount of rainfall recorded when it is present is never very great.

For the sake of easy comparison, 1 in of rain is equivalent to 25 mm, and, spread over an acre of ground, this amounts to just over 100 tons of water. A 'rainday' is any 24-hour period, normally commencing at 0900 GMT on which 0·01 in or 0·2 mm or more of rainfall is recorded.

Rainfall and other cloud-borne precipitation will be described as 'continuous' if it occurs for periods of several hours' duration. 'Intermittent rain' means rainfall that is not continuous over a considerable period of time: for example, for several non-consecutive hours or half-hours of one day. But the varying periods of rainfall may be of quite substantial duration when added together for the 24-hour period, and the sky remains overcast during the intervals when rain is not actually falling.

Then there is the term 'occasional rainfall', which implies that, despite the overcast or near overcast, periods of rainfall are relatively short in duration and occupy only a small fraction of the total time. By contrast, the term 'shower' means a brief fall of precipitation, with breaks in the clouds after the clearance.

When the sun shines upon raindrops, whether the observer is just a few yards or a distance of several miles from the rain, a rainbow is seen. It is caused by the sunlight being bent as it passes through the raindrops, revealing its different spectral colours. The coloration of each rainbow depends on the size of the raindrops. When drops are more than 1 mm in diameter rainbows are brilliant and the limiting colour distinctly red.

With drops of 0·3 mm in diameter the limiting colour is orange, while inside the violet there are bands in which pink predominates.

With smaller raindrops secondary bows appear separately from the primary one, while, with very small drops of about 0·05 mm in diameter the rainfall degenerates into a white fog bow with faint traces of colour at the edges. Rainbows are not infrequently observed by moonlight, but because the human eye cannot distinguish colour with faint lights the lunar rainbow appears to be white.

Moisture inside clouds that have temperatures below freezing point takes the form of ice crystals, which may grow in size as condensation increases or by colliding and joining with each other, as raindrops do. They then become snowflakes and those that reach the ground may be very large. Some have been known to be over 4 in across. On analysis under the microscope, snowflakes have been shown to have complicated structures, all very beautiful, ranging from needles to six-pointed stars with numerous spreading branches.

Fig. 1. Snowflakes.

A foot in depth of freshly fallen snow is roughly equivalent to 1 in of rain, but the ratio can vary considerably according to the total depth and texture of the snow, for after it is deposited it will undergo various changes in texture, and therefore in density, particularly at the lower levels.

A day of snow in British meteorology means any 24-hour period ending at midnight GMT upon which snow is observed to fall, no matter whether the amount is small or great. There is also the term 'snow lying' which is used when, at the time a particular weather observation is made, half or more of the nearby ground is covered with snow. This ground is defined as 'the flat land easily visible from the station and not differing from it in altitude by more than 100 ft'.

During our colder weather spells one frequently encounters snow drifts near high ground, which invariably cause serious dislocation of traffic. Drifts form because when snow falls on a day of strong wind it does not readily settle in open places but rather in regions that are sheltered from the full force of the wind. Drifts can also be caused by the action of the wind upon snow that has already fallen, and at the time the sky may be quite clear of clouds.

The term 'blizzard' is one that has come to be applied to any high wind that accompanies drifted or fallen snow, but originally it applied to the intensely cold north-westerly gales, accompanied by fine drifting snow, that crossed much of the northern states of the USA in winter.

Sometimes snow and rain will fall together, or snow will partially melt as it travels towards the ground. The precipitation will then be termed 'sleet'.

Pellets of ice that fall from cloud to ground are known as 'hail'. They are usually hard, partly transparent, but collectively they look white. Hail falls only from *cumulonimbus* clouds. The hail pellets are formed from raindrops that are thrown rapidly upwards by cloud currents fierce enough to tear an aircraft to pieces. On reaching the cold levels in the upper atmosphere the drops are instantly frozen and then transferred by strong downcurrents to warmer levels, where they begin to melt and attract new moisture or collide with raindrops. In either event they will increase in size. Within a very short time the partly melted pellets are thrown to a high level to be instantly refrozen. The process may be repeated many times before the pellets, now hailstones, eventually fall to the ground. Some have been known to grow as large as grapefruit and weigh over 2 lb each. But the larger hailstones occur mainly in tropical countries or over large continents that experience very hot spells in summer, for this is where *cumulonimbus* clouds reach their greatest proportions.

In addition to hail itself, meteorologists recognise the existence of 'soft hail', which consists of white opaque grains 2–5 mm in diameter having a snow-like structure and occurring at temperatures around freezing point—often before or together with ordinary snow. 'Small hail' is a further variety of hail pellet which is semi-transparent; it has a soft hail centre but is covered by a thin layer of clear ice. As with other types of hail, it falls from *cumulonimbus* clouds but is associated with rain rather than snow.

Like hail, thunder and lightning are products of the *cumulonimbus* cloud, which apart from its other characteristics contains large and powerful areas of positive and negative electricity. Lightning is the flash of an electrical discharge between two clouds or between a cloud and the earth. A distinction is drawn between 'forked' lightning, in which the path of the actual discharge is visible, and 'sheet' lightning, which is the flash of illuminated clouds caused by the light of a discharge of which the actual path is not visible.

Photographic research shows that a forked-lightning flash begins with a faint light travelling down from the cloud and leaving a number of branches. Eventually one of the branches approaches the earth. Then a much brighter and more vigorous illumination travels along this path, reaching the earth, and unseen to the human eye travels back towards the cloud. This tends to light up the original branches. Subsequent strokes may have fewer branches or none at all.

Occasionally, during electrical storms, and usually when there is little wind, a phenomenon known as 'ball lightning' may be observed, and during most years there are some instances reported within the United Kingdom. Normally the balls of lightning are between 4 in and 8 in in diameter. Sometimes they occur immediately after a brilliant flash of forked lightning but at other times when there is no flash at all. They may develop during a period of heavy thundery rainfall, but in some places when there has been no rain for several minutes. Their exact composition remains a mystery, and it is not clear how they are propelled and why they have the effects that they do. A number have been observed to drift through the air and vanish harmlessly, but at other times they will enter buildings, even penetrating window panes, and then explode.

The passage of a stroke of lightning to the lower atmosphere causes rapid heating and consequent expansion of the air near to it, followed by an equally rapid cooling and contraction of this air. Thunder is the term given to the resulting noise vibrations. The distance of the lightning flash may be roughly estimated from the interval that elapses between seeing the flash and hearing the thunder, counting 1 mile for every 5 sec.

Continuous thunder for a long period may be explained by the fact that the sound is travelling along different wavelengths and in any case may have varying distances to travel to reach the observer. Also what appears to the eye to be a single flash of lightning may in fact be several flashes all occurring within a second or less along a similar path. Normally thunder cannot be heard unless it is within 10 miles of the observer.

The term 'thunderstorm' indicates thunder and lightning which occur continually over a long period and which are heavy and frequent at times. There is usually heavy precipitation in the form of rain and hail, and, in the case of winter thunderstorms, of rain, hail, snow, and sleet. A 'thundery shower', on the other hand, is a heavy shower of rain or other precipitation accompanied by thunder and lasting not more than 30 minutes. 'Thundery rain' is the type of rainfall which, whether 'continuous', 'intermittent', or 'occasional', is generally heavy when atmospheric conditions are oppressive and close. The rainfall is generally accompanied by thunder.

Sometimes, during very disturbed thundery weather, a funnel or cone will form underneath a *cumulonimbus* storm cloud and descend to the ground. It looks very much like an elephant's trunk and is known as a tornado or, if it occurs at sea, as a waterspout—the latter term being somewhat misleading. The cloud cone is a narrow zone of fiercely revolving winds, and at sea it causes a large area of spray near the surface of the water. Much damage is done by these phenomena, particularly in tropical regions, and they are caused in the main by the explosive effect of the intense and sudden decrease of atmospheric pressure between the exterior and the interior of the cone. At 4 pm GMT on 21 April 1968 a British tornado travelled from Coventry north-eastwards towards the village of Barnacle, a distance of approximately 3 miles, damaging a number of buildings in its path. Before the tornado occurred there was a storm of hail, thunder, and lightning. On a previous occasion the Royal Horticultural Society's gardens at Wisley, Surrey, were badly damaged by a tornado; this formed at 3.40 pm GMT about a mile west of the gardens, swept through the fruit collection, brought down trees on the A3 London to Portsmouth road and finished its course at Wisley airfield, where it damaged some hangars.

It is not difficult for the observer new to meteorology to appreciate that clouds can inflict major havoc for they are manifestations of wind and water and by their size, shape, and movement give some warning of what to expect. But just as interesting in their different ways are the more gentle and subtle of the elements. Take dew. It has a mysterious quality and is in fact moisture that is squeezed out from the layer of air lying just above—and in contact with—the ground. The process takes place during the night or the early hours of the morning but only after a warm day and when the sky is clear or almost clear of cloud with little or no wind.

Under these conditions the ground cools rapidly, by losing its heat by radiation into space. It then cools the layer of air just above it, and if this contains enough moisture in the form of invisible water vapour, and the amount of cooling is sufficient, then water droplets resembling a light fall of rain will form on the ground's cold surface. It will 'fall' copiously on stones, grass, and leaves but only lightly on earth, sand, and gravel. If the temperature falls below freezing point, then ice crystals will form in place of the dewdrops and the deposit is then known as 'hoar frost'.

There is little chance of dewfall or hoar frost occurring during cloudy weather, for clouds act as a blanket and prevent the earth from losing too much of its heat by the radiation process. Thus at ground level there is only a minimal loss of temperature. Again, neither dew nor hoar frost will form if the strength of the wind is greater than approximately 5 mph, for the wind in this case will distribute the coldest airs just above the ground over a vertical thickness of some 50 ft or more, and the condensation that results will then be fog.

Fog in fact is simply a cloud at ground level. If temperatures fall to freezing point or below when fog is forming, deposits of white rough ice crystals known as 'rime' will form to windward of exposed objects. The fog will in fact have become a 'freezing fog', which is always much dreaded by motorists as it plays havoc with windscreens. Oddly enough, the actual particles of fog tend to remain in liquid form, despite the low temperatures, until they come into contact with still colder objects near or above ground level.

Rime ice will often form on telegraph poles and on the foliage of trees and bushes, creating a general fairyland effect. 'Glazed frost', which is comparatively rare in Britain but is moderately frequent in the USA, is the name given to the coating of ice caused by 'freezing rain'—rain falling on to ground that is below freezing point from temporarily warmer air levels above. The raindrops freeze immediately they strike the ground. Similarly, tree branches and telegraph wires become heavily coated with ice and may break down under the greatly increased weight.

During cold spells frost may occur without precipitation when strong easterly winds are blowing, and will probably be intense and last, night and day, for a period of several days or even for a week or more. This is known as 'black frost'.

During the winter months the formation of fog and frost over land areas are invariably linked, and the tendency is often for one or two foggy days or nights to be followed by others. There may be some lifting of the fog at times during the daylight hours, but it will tend to persist night and day near towns to an increasing extent as the length of the foggy spell becomes more extended. As more and more counties become affected, the fog layer will increase in depth as well as horizontally. Summer and early autumn fogs, on the other hand, which form over valleys, fields, marshes, and streams, do not extend upwards to any great height and sometimes the tops of trees may remain visible. These fogs clear rapidly during the early morning periods, often lifting to become low *stratus* cloud before dissolving altogether.

A further type of fog occurs at sea and over coastal regions. It is only seldom that land-made fog drifts seawards to become extensive away from the coasts. Normally sea and coastal fogs are formed by comparatively warm winds blowing over a cooler sea surface and causing condensation of the air in the lower levels. The meteorologists call this 'advection fog', which is simply a reference to its horizontal motion. Some sea fogs may penetrate inland for a few miles, and may indeed affect the weather for the whole of a day or more in coastal areas. But away from their influence inland, the sky may be clear and the general daytime weather fine and sunny.

A different classification of fog types is used in official analysis of daily weather to indicate visibility limits. 'Dense fog' means that the visibility is less than 44 yd, 'thick fog' that it is more than 44 yd but less than 220 yd, 'fog' that visibility is 220–440 yd, and 'moderate fog' that it is 440–1,100 yd. A visibility level greater than 1,100 yd but less than 2,200 yd is referred to as 'mist' or 'haze', the latter term being used only when the obscurity of the atmosphere is caused by dust or other solid impurities, and the former when it is caused by atmospheric moisture.

Summer fog or mist may be dispersed by the heating action

of the sun on the ground, leading to the warming of the air above the ground. Wind will also disperse it. But an easterly wind will always tend to maintain hazy conditions in the atmosphere in most areas of Britain. In winter the sun in our own high latitudes is seldom strong enough to burn off a fog layer, so that wind is then the only clearing agent outside artificial measures.

Of all the elements of weather wind is probably the most baffling. It is basically just air in motion. But for the great variety in wind strength and direction that we may experience in the course of even a single season the causes are many and complex. The earth's wind circulation is in the first place geared to the sun's heating of the earth. The earth heats the air above it, and over the equator in particular, where the most constant heat is received; here the air rises vertically. As it rises it becomes cooler and eventually travels horizontally northwards and southwards some 5–8 miles above the earth. It falls to the ground again at higher latitudes, from where it makes its way back towards the equator. However, as winds cross from land to sea, and *vice versa*, the form that they take and the strength at which they blow are affected all the time by constantly changing temperature patterns and by mountains and other topographic features, as well as by the spinning of the earth on its axis. Fig 2 shows the directions taken by the world's prevailing wind systems at surface level.

So far as British weather is concerned, the effects of the winds are often paramount. What is important, apart from the strength and direction of the wind on any particular day, is its 'fetch', which is the distance it has travelled in a particular direction. For as we import winds, which we constantly do, we also import the general weather that they bring with them.

In 1805 wind speeds were graded by Admiral Beaufort, and the scale named after him, and shown on pages 32 and 33, has been in use ever since.

This was first designed for use at sea but was adapted also for land use at a later date. Hurricane-force winds are not common in Britain, but they do occur from time to time—for example, during the mid-January storm of 1968 (see Appendix 3). Gale warnings are issued by the Meteorological Office when an average wind speed of Force 8 or more is expected or if it is thought that, in any event, gusts of wind will reach Force 9 or more.

Strong south-west winds may bring mild temperatures, even in mid-winter, but with winds from any direction temperatures are always comparatively higher with a light wind than with a stronger one.

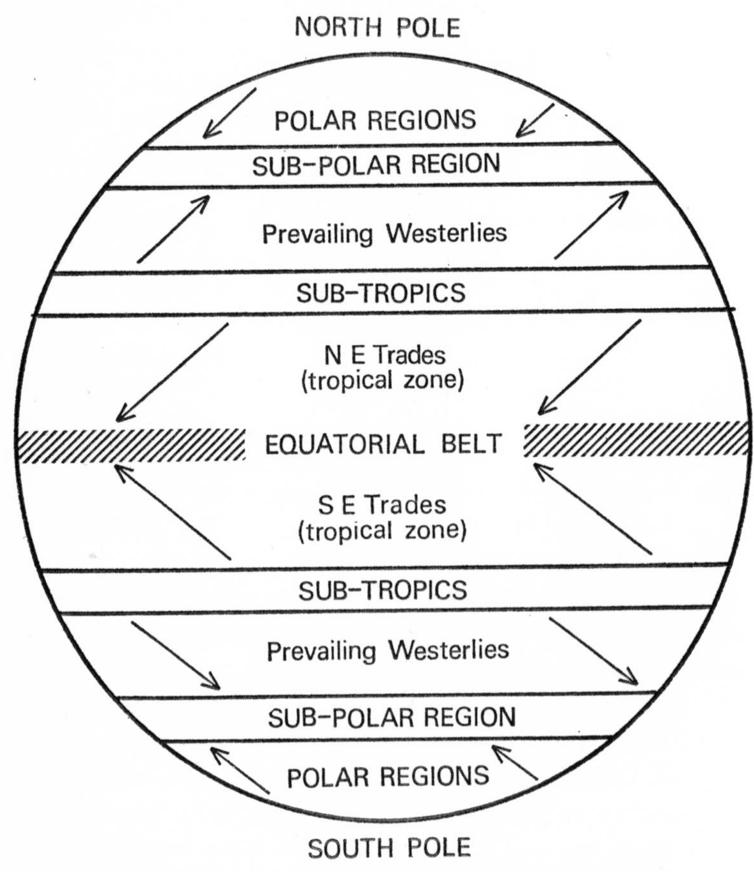

Fig. 2. Prevailing winds over the globe at surface level.

The foreign air masses brought to Britain as a result of the wind's travel have been closely studied and have been categorised as follows:

1. *Arctic and Polar Continental*
   Winds blow from Greenland, northern Russia, Finland and Spitzbergen. They bring mainly dry, clear, cold and sunny weather in winter but tend to be rather showery and cold in summer.

2. *Polar Maritime*
   Winds blow from the direction of west coast of N. America and bring cool weather, with frequent showers, some of which are thundery. Inland and east coast areas of Britain suffer least from the showers and experience long sunny periods, particularly during early mornings and late afternoons. The winds are typical of the unsettled spells of mid-spring, when the general seasonal temperature rise is temporarily checked or reversed.

3. *Tropical Continental*
   Winds blow from due south bringing air from southern Europe and occasionally from North Africa. Although they are comparatively rare over Britain, an example occurred on 1 July 1968 when a flow of air direct from the Sahara brought fine red dust to many parts of England. The winds, which are basically very dry and are comparatively warm even in winter, may mix with Atlantic airflows on approaching the English Channel, to produce thunderstorms or thundery rainfall.

4. *Tropical Maritime*
   Winds blow from the southern part of the North Atlantic Ocean. They bring fine, mainly warm weather during all seasons if they originate from the Azores when barometric pressure over Britain is high. But if pressure over Britain is low or falling, they bring mainly damp and certainly very cloudy conditions, with light rain or drizzle

during the winter when temperatures are mild. Sea fogs are also common, particularly in winter, when they may be very persistent near southern and western coasts of Britain; they can also occur during the summer months but do not often penetrate very far inland.

These basic air types that reach the British Isles will be modified from time to time. For example, a south-easterly wind can bring either tropical continental or polar continental air, according to the fetch. In winter the latter type of air is more common than the former. A wind from due west will bring air in which either the polar maritime or the polar continental influence predominates, again according to the fetch.

BEAUFORT SCALE

Beaufort Number	*Explanatory Titles*	*Limits of Speed (33 ft) in the open MPH*	*Specification of Beaufort Scale for Use on Land based on Observations made at Land Stations*	*Specification of the Beaufort Scale for Coast Use*
0	Calm	Less than 1	Calm; smoke rises vertically.	Sea like a mirror; no ripples.
1	Light air	1–3	Direction of wind shown by smoke drift, but not by wind vanes.	Small scale-like ripples.
2	Light breeze	4–7	Wind felt on face; leaves rustle; ordinary vane moved by wind.	Small wavelets; glassy crests.
3	Gentle breeze	8–12	Leaves and small twigs in constant motion; wind extends light flag.	Large wavelets; crests begin to break; some 'white horses'.
4	Moderate breeze	13–18	Raises dust and loose paper; small branches are moved.	Waves become longer. Many 'white horses'.

5	Fresh breeze	19–24	Small trees in leaf begin to sway; crested wavelets form on inland waters.	Moderate waves with pronounced length; many white horses; some spray.
6	Strong breeze	25–31	Large branches in motion; whistling heard in telegraph wires; umbrellas used with difficulty.	Large waves, with extensive foaming crests.
7	Moderate gale	32–38	Whole trees in motion; inconvenience felt when walking against wind.	Sea heaps up. Foam begins to blow off in streaks.
8	Fresh gale	39–46	Breaks twigs off trees; generally impedes progress.	Moderately high waves, with much foam.
9	Strong gale	47–54	Slight structural damage occurs (chimney pots and slates removed).	High waves, with dense foam, sometimes affecting visibility.
10	Whole gale	55–63	Seldom experienced inland; trees uprooted; considerable structural damage occurs.	Very high waves with long overhanging crests. Great foam patches. Surface of sea white.
11	Storm	64–75	Very rarely experienced; accompanied by wide-spread damage.	Waves so high that ships in troughs cannot be seen.
12	Hurricane	Above 75	—	Air filled with foam and spray.

3

THE METEOROLOGISTS

THE Meteorological Office was established in 1855. Its function at that time was to provide information purely for seamen, but since then a vast amount of intricate data relating to weather conditions all over the world has been accumulated. The Met Office—as it is now more familiarly known—is a department of the Ministry of Defence and is divided into several branches or divisions—Central Forecasting, which provides the weather forecasts for the public; Upper Air Branch, which is concerned with atmospheric research of all kinds; Marine Branch, which serves shipping interests exclusively; and so on.

But there is nothing novel or modern in man's attempt to predict the weather. Scorched by sun, nipped by frost, and constantly harassed by wind, drought, and flood, primitive man set himself the task of trying to understand the elements, for such knowledge was essential to his survival: 3,000 years ago the Chinese, clustered along the fertile banks of the Yellow River, were using the stars to foretell future weather; and the Babylonians and Chaldeans were studying thunder in relation to the phases of the moon.

But mixed with the attempts at forecasting weather was a great deal of superstition and, at times, fatalism. In Greek mythology Zeus unleashed a favourable wind in return for an appropriate offering. Other gods performed services of a like nature, and Poseidon would raise a whirlwind to defeat an enemy. The point was that if a storm or any other meteorological event could be readily explained, then it would have

Page 35: *(above)* Underside of a large *cumulonimbus* shower cloud, with rainbow to windward indicating further showers to come; *(right)* distant thunderheads; *(below)* *cumulonimbus* with typical anvil top.

Page 36: *(top) Mammatus* indicating downdraughts inside thundercloud; *(centre)* start of a squally day, Padstow, Cornwall, 8.30 a.m., February 1967; *(bottom)* same location ten minutes later.

to be attributed to a divine power, and this trend of thought reveals itself in many of the early Hebrew writings. Job declares:

> He draweth up the drops of water which distil in rain from His vapour; which the skies pour down and drop upon man abundantly . . . Great things doeth He, which we cannot comprehend. For He saith to the snow, fall thou on the earth; likewise to the shower of rain, and to the showers of His mighty rain . . .

But even from the times of the ancient Greeks a more materialistic school of thought began to develop. In his comedy *The Clouds*, written in the fifth century BC, Aristophanes staged an argument as to whether thunder was made by Zeus or by an inevitable clash of clouds in a vortex of air. The scientific approach to meteorology had firmly taken root by the time of Aristotle, in whose works concerning the physical processes of the atmosphere there was hardly a reference to the supernatural. Aristotle might indeed claim the distinction of being the father of meteorology, for in his classic work *Meteorologica* he dealt with the formation of most of the elements of weather with great objectivity, and his pupil Theophrastus carried the work a stage further. But in those days there were no instruments to measure temperature, rainfall, humidity, or atmospheric pressure, and for many years the early theories on weather and climate were destined to remain unverified.

Not, indeed, until the seventeenth century was anything further added to the world's general knowledge of the weather. But weather lore and astrology were continually practised. A wealth of observations was collected regarding the lower forms of life, for it was widely believed that plants and insects were sufficiently attuned to the weather to give mankind advance warning of what was to come. A massive treatise on sky and weather signs had been compiled by Theophrastus in the fourth century BC and, later, men like the Venerable Bede and Roger Bacon wrote equally great volumes on weather forecasting that revived what the ancients had observed previously, often many centuries before, but which did not add greatly to current knowledge.

But in the twelfth century William of Conches recognised that many of the weather-making processes took place in the upper atmosphere well above the surface of the earth and that winds, weather and ocean currents were inter-related. What was more, weather conditions appeared to be greatly modified by the presence of mountains and lakes. But it was not until the following century that the first weather logs of any consequence found their place in society. A notable example was William Merle's *Observations of the Weather*, which covered the years 1337–44, the object of which was to record all abnormal weather events and to provide material on which to base forecasts of future conditions.

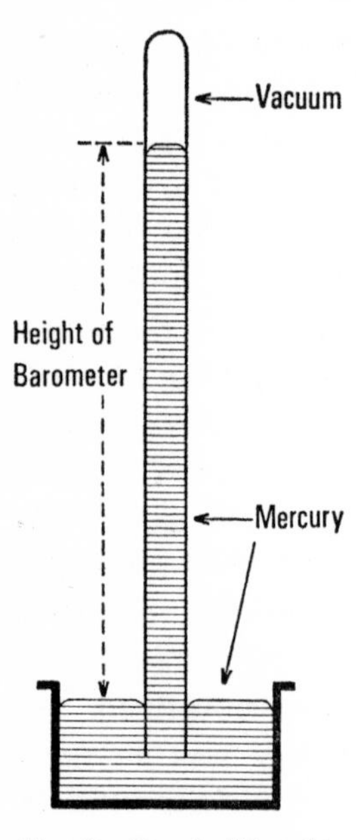

Fig. 3. Torricelli's tube barometer.

There was no real widening of the horizon until 1643, when the barometer was invented to measure atmospheric pressure. Torricelli, who was Galileo's assistant, found that the earth's atmosphere had weight. In fact, it was heavy enough to support a column of mercury about 30 in high. The liquid was put into a tube that was closed at one end. The other end was inverted into a small open cistern, and the weight of the atmosphere on the mercury in the cistern supported the weight of the liquid in the inverted tube. However, its height in the tube was subject to variations from day to day, even at times from hour to hour, and these were found to be associated with changes in the weather.

In 1670 Robert Hooke produced the first 'wheel' barometer with the word 'Change' at the 29·5 in mark. On the side where the mercury reached its higher levels were the words 'Fair', 'Set Fair', and 'Very Dry'; and on the other side were the words 'Rain', 'Much Rain,' and 'Stormy'.

The seventeenth century also saw the invention of two other instruments. It was shown by Galileo that the expansion and contraction of a liquid in a glass vessel was matched by changes in the warmth or coldness of the atmosphere. Alcohol or mercury was enclosed in a glass bulb that led directly into a long tube of fine bore, and the imprisoned liquid could move along this tube as it was cooled or warmed by the temperature of the air. And so the thermometer came into being. The other new instrument, which was soon considered to be necessary for a proper understanding of the weather processes, was the hygrometer. This comprised a length of human hair, fixed at one end but free to move at the other, that was attached to a needle carrying a pointer. The hair contracted during dry weather and expanded when it was moist. In fact, modern hygrometers work on exactly the same principle. Freed from natural oils, hair increases in length by about $2\frac{1}{2}$ per cent from a completely dry to a completely wet state.

The barometer was, correctly enough, considered to be the premier instrument, but too much faith was placed in it. It was regarded by many as being capable of solving all weather problems, an attitude that still persists today in some quarters. Scientists, on the other hand, soon realised that the connection between barometric levels and weather changes was by no means a simple one; in fact, at times, there was no direct connection at all. It became clear that the movement of the mercury over a period of time was more important than the height of the barometer at a given instant. It was also evident that much more information about the weather was required before forecasting could be carried out with any chance of success.

This brought progress to a halt. The eighteenth century concerned itself very largely with theorising. But although everyday weather changes were to remain a mystery for many years, it so happened that the opening up of the world's frontiers and oceans gave men a rudimentary knowledge of climate, and in 1688 Edmund Halley (of comet fame) made a diagram of trade winds and monsoons.

Early in the nineteenth century it came to be seen that weather was very much a moving phenomenon and that warnings of travelling storms might be given if communications systems could only be improved. Weather observations were taken much more regularly and with greater precision. What is more, the public interest in them was growing. The invention of the electric telegraph, a device for transmitting messages between two distant points, was the vital step forward. The invention was first patented in 1836 and in the following year it was tried successfully between Euston and Camden Town stations.

The credit for adapting the telegraph to the needs of meteorology went to James Glaisher, FRS, who was a member of the staff of Greenwich Observatory, and to the *Daily News* who commissioned him to organise a system of weather observations that were to be made each day at 9 am and telegraphed to London for publication in next day's issue of the newspaper. These observations concerned themselves with the strength and direction of the wind and with the general state of the weather. In the USA Professor Henry of the Smithsonian Institution realised the possibility of using the electric telegraph for collecting current information on the weather at almost the some time as Glaisher.

In 1851 came a further step forward, when charts based on telegraphed reports and produced by lithography were sold at the Great Exhibition in Hyde Park to astonished visitors. The first chart made its appearance on 8 August. It showed the direction of the wind at each station by means of an arrow, the barometric pressure (calculated in inches and hundredths of an inch), and the general weather, which was denoted by means of various key letters (*f* for fine, *c* for cloudy, *r* for rain, and *s* for snow). But no forecast was attached to the chart, it was purely a representation of actual conditions.

Two years later a special maritime conference was called, for the subject was now attracting official attention and weather information was much needed by ships at sea. In 1855 it was decided to create a Meteorological Department at the Board of Trade. In charge of it was Admiral Fitzroy, who set himself the immediate task of compiling instructions for the interpretation of barometric readings, taking into account the rapidity of each change coupled with the direction and strength of the wind.

In 1859 Fitzroy was asked to organise the collection of weather reports by telegraphy and to issue warnings of severe gales based upon them. In Paris a similar type of service was being organised, and it was arranged between Britain and France in 1860 that an exchange of information should take place. It is interesting to note that not all things change in a quickly changing world, for Fitzroy's method for using the gale warnings is still in use today. It provides for canvas cones to be hoisted at points on the coast as visible signals of expected gales. The cone is hoisted point downwards for a southerly gale and point upwards for a northerly gale.[1]

In 1861 daily forecasts were issued as distinct from storm warnings, but both were discontinued in 1867 when the supervision of the Meteorological Department was transferred from the Board of Trade to the Royal Society. The Board of Trade protested and eventually, in 1874, the service was resumed.

The first phase in the development of an official forecasting service had now been completed. The principle of basing the prediction of future weather on the revelations of plotted charts had been acquired and approved and the basic organisation to deal with them was in being. Ahead lay expansion, but it was some years before the pace became anything like hectic.

At first the emphasis was on the improvement of communications. The network of submarine cables over the globe enabled the forecasters to extend the area covered by their weather charts until these included the whole of Europe and the outlying islands of the Faroes, Iceland, and the Azores. Then came the very widespread application of a new system of communications. Marconi constructed a workable wireless signalling apparatus in 1895, and by the early years of the twentieth century its application was very wide. Meteorologists now could obtain observations from Jan Mayen and Spitzbergen in the Arctic and from ships at sea. The same device made it possible for nations to exchange weather reports quickly and economically, and as radio techniques improved it became feasible to use specially designed equipment carried aloft by balloons to measure weather conditions up to comparatively great heights in the atmosphere. Distant thunderstorms could also be located.

And here, one might say, ended the second stage in the weather forecasting saga. The first stage had been forced upon our ancestors by the needs of mariners. The next stage, with the marine interests still the main consideration, had crossed into the new century and had converted the forecasting organisation, now the Meteorological Office, into a national institution from what, before, was purely an experiment.

During the third stage, which takes us to the present day, the meteorologists have made further progress as a result of developments in the air which have led to greater all-round knowledge. But although aviation has helped the forecasters by providing them with extra observations, it has also passed to them increasing responsibilities. It is fortunate that, for most of the time, practical forecasting techniques have kept pace with improved methods of observation, though until the First World War we had little knowledge of the physical processes taking place in the atmosphere. Nevertheless, forecasts were based—as they are to a large extent today—on the movement towards (or the retreat from) Britain of weather systems of varying intensity. The association of types of weather with barometric movements became more widely understood, and also the changes they brought in wind direction and strength and, therefore, in temperature.

Between 1914 and 1918 important research carried out by a number of Norwegian meteorologists gave the science of weather forecasting a 'new look'. The research proved that the daily movements of weather systems of high and low barometric pressure were not the prime causes of weather changes; these were caused by the large-scale movements of air masses which, although they may travel great distances from their places of origin, tend to preserve many of their original characteristics and do not mix very easily. In order to forecast, therefore, it was essential to concentrate on the air masses in the first place and, secondly, on the lines or 'fronts' which divide them.

This linked up very well with previous forecasting techniques, since it was found that the fronts were common in most low-pressure areas—called 'depressions' or just 'lows' for short—and marked the change from one air mass to another.

Between the two world wars the development of the RAF and civil aviation created the need for a large number of forecasting centres to deal with pilot briefings. During the Second World War and, more particularly, since then, these have been given in increasing detail. Flight ranges, both vertical and horizontal, were extended, and, as it became particularly important for aircraft to fly to schedule, the forecasting of wind strength at various levels of the atmosphere needed to be exact. Visibility, cloud cover, and general weather conditions at points of ascent and descent were also given special attention.

So far as international co-operation was concerned, meteorology became a definite trend-setter. As long ago as 1873 a world congress of meteorologists was held in Vienna for the purpose of standardising different types of weather stations, the equipment they should carry, and the duties they should perform. Five years later a special International Meteorological Organisation was formed, and this met periodically to regulate the exchange of information between the nations. Special international codes were agreed upon so that translations of weather reports from one language to another were unnecessary.

By the 1930s weather forecasters throughout Europe were able to know the existing conditions at any station from the Arctic to northern Africa and from the American continent to central Russia. Only during wartime did the collection of weather reports become dependent on national or political boundaries, and, today, international co-operation in this field is going ahead at a remarkable pace. The regulating authority is now the World Meteorological Organisation, which is a branch of the United Nations Organisation at Geneva (see Appendix 4). Here frequent conferences of technical experts from all parts of the world promote meteorological studies. In addition—and this is a distinct post-war trend—they examine practical ways in which information can be adapted to meet requirements in many fields of human endeavour. In Britain the Meteorological Office is becoming increasingly alive to new opportunities. After the last war it created a special Agricultural Branch and began a drive to improve the forcasting services given to the general public. Its efforts, since then, in the direction of expansion and improvement have been considerable, particularly in recent years (see Chapter 5). The modern meteorologist is having to face challenges that are as great as those of the past, and many young people interested in the natural sciences find that meteorology offers an interesting career with scope for personal initiative.

The Met Office, which is a part of the civil service and is financed by the Ministry of Defence (formerly by the Air Ministry), gives instruction in weather forecasting to accepted school leavers who, at the age of 16, possess a GCE at Ordinary Level in four subjects, including English language and mathematics or physics, or in general science. These entrants become Scientific Assistants, whose main duty it is to take regular observations of all the weather elements. Pensionable posts are available after a period of one year.

Those who wish to become forecasters in the Met Office are

appointed, after examination, to Experimental Officer grade, and entrants require either a university degree or GCE passes at Advanced Level in physics or mathematics, or in two mathematical subjects and an Ordinary Level pass in physics. Apart from being given forecasting duties, Experimental Officers assist in research and may have exceptional opportunities to travel. Postings are made to places such as Gibraltar, Malta, Singapore, East and North Africa, and the Falkland Islands. Sometimes the officers in this grade are called upon to take command of local meteorological offices.

Of a staff of some 4,000, approximately 170 men in the Met Office are in the highest class of the Scientific Civil Service, which is the Scientific Officer class. Around seven are chosen each year, and, to qualify, an applicant must have obtained first or second class honours in a scientific degree, and mathematics and physics are the most suitable subjects. A Chief Scientific Officer commands a salary of just under £5,000 per year. At the head of the organisation is the Director-General and a number of Deputy and Assistant Directors who are elected from the Chief Scientific Officer grade.

4

THE WEATHER CHARTS

This chart shows the weather situation over the British Isles at noon today. As you will see, the winds are strong, and the heavy showers will continue in all districts. Tomorrow, however, the picture should look something like this—with the showers dying out gradually and perhaps disappearing altogether in the East by the end of the day. In northern districts the showers will be heavy for much of the day over exposed high ground and will produce hail or even snowfall at times. Due to the north-westerly airflow, temperatures will be mainly below average in all areas. And now, here is the latest Atlantic weather chart . . .

FAMILIAR words indeed! The British are used to bad weather though dry periods with sunny spells are not so uncommon as some would suppose. And they are accustomed to the appearance of the Met Office forecasters on the television screen and to the charts that they display, showing actual and expected future conditions. But little, if anything, is said about how these charts are constructed and how the various systems marked on them produce the changes in weather that they do. Not that this is secret information by any means, except in wartime, but the fact is that time prevents the forecasters from doing much more than outlining the main features of the latest national and regional forecast bulletins, and even these have to be abbreviated to some extent and occasionally read at fairly high speed.

Yet if we can discover how the forecaster draws his conclusions, this will lead to a better understanding of each current weather situation and of what to expect in the near future. It is also helpful when we come to interpret forecast bulletins for personal use in our own local areas.

In order to produce a national forecast in the UK it is necessary to refer to charts that show not only the British Isles but also the surrounding areas of ocean and continent from which weather may travel towards us. In practice it is found that weather reports stating existing weather conditions must be received from a close network of stations over the whole of Europe, the North Atlantic, and the Mediterranean at least every 6 hours and preferably every 3 hours in order to make satisfactory forecasting charts for Britain for periods of 12–24 hours ahead. To give a reasonably accurate outlook for a further 24 hours it may be necessary during periods of unsettled weather to extend the network as far west as the North American coast, possibly eastwards well into Asia, and southwards to North Africa. Detailed and highly accurate forecasts for aviation use are based on hourly weather reports from stations within the areas crossed by the various air routes.

The flow of weather reports reaching the Central Forecasting Office at Bracknell, Berks, where the Meteorological Office now has its headquarters, increases every year; and today a large proportion of these are fed into computers which carry out mathematical sums and save the forecasters valuable time. They are even used to produce short-term upper wind forecasts for fast-flying aircraft. But, as yet, machines cannot produce generalised forecasts on their own. Trained human forecasters are still needed for this task, and the basic method of producing the forecasts has not changed over the years.

This method involves the use of charts—'synoptic' charts, as they are called, because they are built up from observations taken simultaneously over large areas at the same time of day, and the most familiar of these are known as 'surface' charts because, as distinct from the special upper atmosphere charts that are also made, they deal exclusively with observations taken at ground or at sea level. Before they are used, the charts depict the British Isles and the surrounding land and sea areas in outline and show a small blank circle for each land weather station or fixed ocean weather ship station.

As soon as they are received, the observations are plotted on to the charts over the appropriate station circles for 3 am and 6 am GMT and for each subsequent 3-hour period. The first of these arrive from Britain and nearby continental stations, then come a number of Atlantic reports, and, finally, the remainder of the station reports and a number from more distant regions such as N. America, Asia and N. Africa.

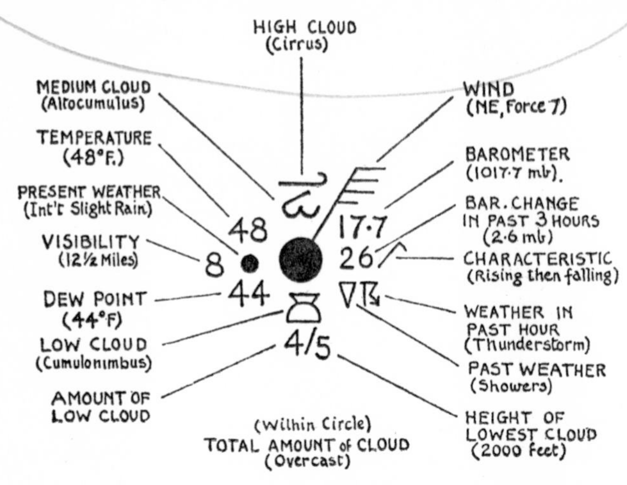

Fig. 4. How station reports are plotted on a weather chart.

Another operation involves the plotting every hour of large-scale charts that show the weather over the British Isles in greater detail.

A considerable number of the reports received at the Central Forecasting Office come from fully equipped and fully staffed stations that provide surface observations of all kinds, and very often these are situated at airfields and carry out upper atmosphere probes as well. Other reports come from auxiliary stations, where less comprehensive observations are made. Some of these stations are at airfields, though a few of them are manned by a single local observer who works for the Met Office on a part-time basis. One such observer has been a postmistress at a remote village in the Pennines. From a few

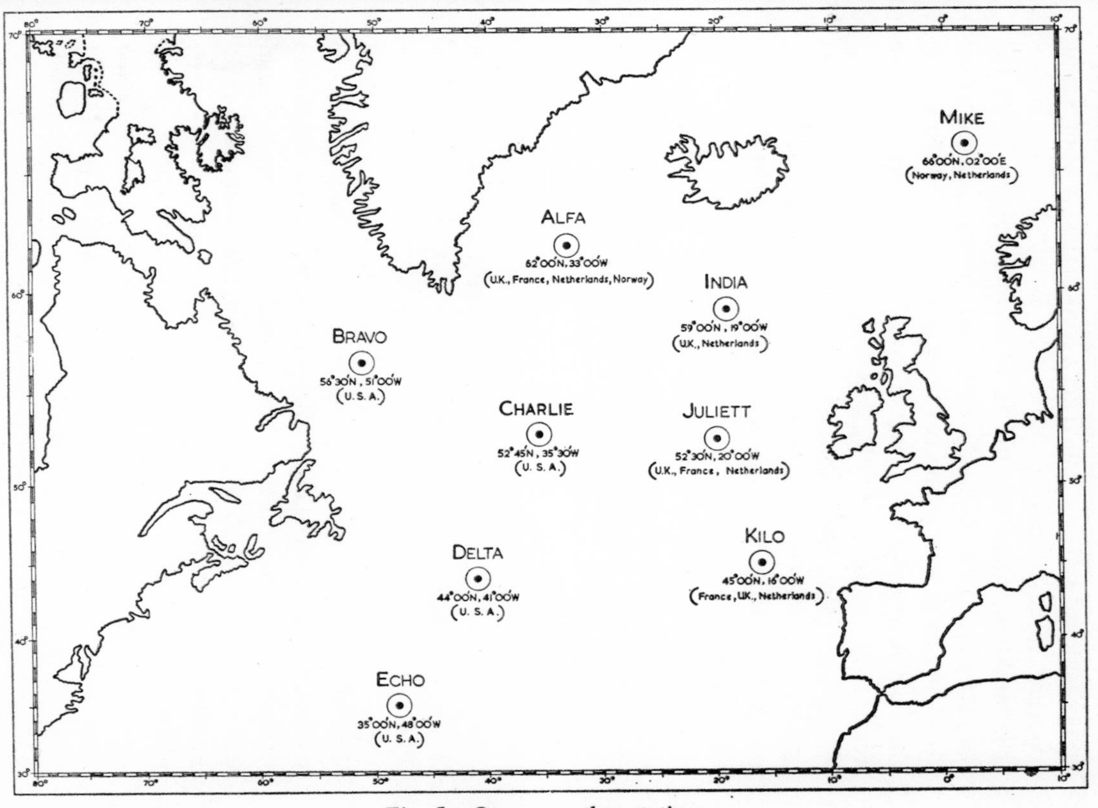

Fig. 5. Ocean weather stations.

coastal areas that are of importance in forecasting but which are far from main weather stations, use is made of lighthouses and coastguard stations, which are asked to send in abbreviated reports at certain hours.

In addition, there are the fully oceanic reports from the weather ships, four of which are British, by ships of the Royal Navy, and from a large number of passenger and freight vessels under an international scheme of voluntary co-operation. The weather-ship reports are fully detailed and have the advantage of coming from fixed stations. This is not the case with reports from merchant ships which, useful as they are, do not always come from those parts of the ocean where information is most needed.

Yet over 700 British ships alone take part in this scheme and altogether about eighty reports are received each day from ships of various nations. The programme of observations is broadly similar to that of land stations, but there are some modifications, such as the omission of the amount of rainfall. This cannot at present be accurately measured at sea by conventional methods due to the disturbance of the airflow by the ship if it is moving, and, in any case, by the effect of spray. On the other hand, ships can provide valuable information, for example, sea temperatures and wave and swell observations. Unfortunately, some of the observations are inclined to be rather sketchy, for much depends on the duties of each voluntary observer, and it has therefore been found necessary to classify these ships reports into several categories: those from 'selected' ships which are able to make full reports—normally at midnight, 6 am, noon and 6 pm GMT; those from 'supplementary' ships, which report less fully but at the same times daily; and, finally, those from 'auxiliary' ships—mainly coasting vessels, lightships and distant water trawlers—which send in occasional reports of a non-scheduled nature.

Each full weather observation yields a large amount of information, and it is the observer's duty to convert this into a coded report for onward transmission through the land telegraph line or, if at sea, by radio telegraph. This takes the form of eight groups of five figures each and ensures that a great amount of information is packed into a comparatively short message.

At the Central Forecasting Office the messages are simultaneously decoded and recoded as they are plotted on to the

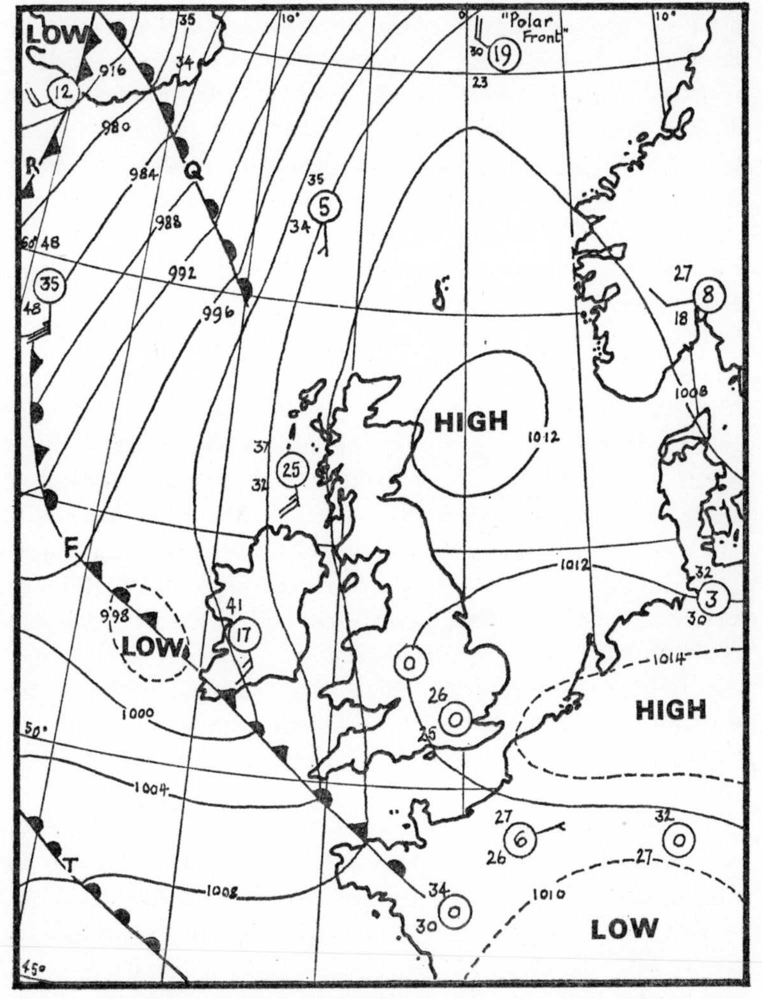

Fig. 6. Simplified weather chart; wind speeds given inside station circles.

appropriate station circles on the current charts, and this task can now be carried out automatically by computerised machinery. For their part the forecasters can grasp the full meaning of most of the plotted weather-station reports without having to make more than occasional reference to the appropriate codes.

When each chart is fully plotted the many land and sea stations, with their combined information, will appear like stars in a mosaic, and the forecasters must take immediate steps to obtain a picture of the weather situation as a whole. This is done by drawing lines on the chart to connect those places with the same barometric pressure. These are called 'isobars' and are equivalent to the contour lines on an ordinary relief map, only instead of showing regions of high and low ground they pinpoint the respective regions of high and low atmospheric pressure as measured by the barometer.

The relief map and the weather chart have another point of comparison. When the contour lines on the former run close together the gradient is steep and when they are far apart the ground slopes only gently. Similarly, on the weather chart, when the isobars are close together the pressure gradient is steep and, as a result, winds blow strongly. But when the isobars are widely spaced the pressure gradient is referred to as 'slack' and winds are light or else there is calm. The chief difference between the features on the relief map and those on the weather chart are that the latter are normally on a much larger scale. An isobaric hill, representing a region of high barometric pressure, or an isobaric hollow, representing a region of low barometric pressure, may cover an area as large as the whole of Europe.

However, isobars are not the only lines that appear on the weather charts. Thicker notched lines are also drawn in to represent the fronts which are the boundary lines between the different air masses. There are really only two basic types of front: first, the 'warm' front, where a warm airstream is pushing itself against a colder one; and secondly, a 'cold' front,

Page 53: *(above)* Lens-shaped *stratocumulus* formed by the spreading out of decaying *cumulus* at the end of a showery day; *(below left)* small break in centre of shower clouds; the weather will remain unsettled; *(below right) altocumulus*: this type of curdly sky indicates changeable weather but no immediate rain.

Page 54: *(above)* Hooked *cirrus* merging into *cirrostratus* sheet on the right, warning of the approach of an Atlantic depression; *(below)* thickening *alto-stratus* sheet. Rain from depression will fall within a few hours.

where a cold airflow is driving its way under a warmer one. An 'occluded' front occurs where the cold front has caught up the warm front, and here, at surface level, one cool or cold airstream is being replaced by another of somewhat different origin.

When the time comes to make a forecast, more than one weather chart is required. It is necessary to compare the most recent one with the several previous charts so as to determine how the various air masses and the relevant high- and low-pressure systems have moved. Calculations are then made of the speed and intensity of all the moving weather systems, but what these charts do not show is whether the movements will continue at the same rate and vigour as before or whether they will accelerate, slow down, intensify, or become weaker. A particular feature might appear quite clearly on three successive charts but then change direction or disappear. What is more, by the time the next weather chart is plotted and drawn up, entirely—one hopes, not unexpected—new features might reveal themselves.

As each chart is produced, forecasts must be made according to current schedules by estimating how the existing weather situations will develop during the next 12–24 hours. The future sequences are deduced by methods that are partly scientific and partly empirical—that is to say, they rely on past experience to some extent. At the Central Forecasting Office this process is carried out by a group of senior forecasters rather than by individuals on their own, but mistakes can occur for the simple reason that there are still some gaps in our knowledge about the physical behaviour of the atmosphere. It must be remembered, too, that the confidence of the forecasters in what they pronounce will vary according to the completeness of the information in front of them on the charts and also, in large part, to the type of situation that the chart reveals. Some are much easier to interpret than others in the light of present-day knowledge.

So, despite the marvels of the modern weather charts and

the experience gained from many years of practical work, the meteorologists have to admit, however reluctantly, that forecasting is not a exact science. Perhaps it never will be, but the aim is to make it so, and all modern techniques are designed with this in mind. To cut guesswork and 'rule of thumb'

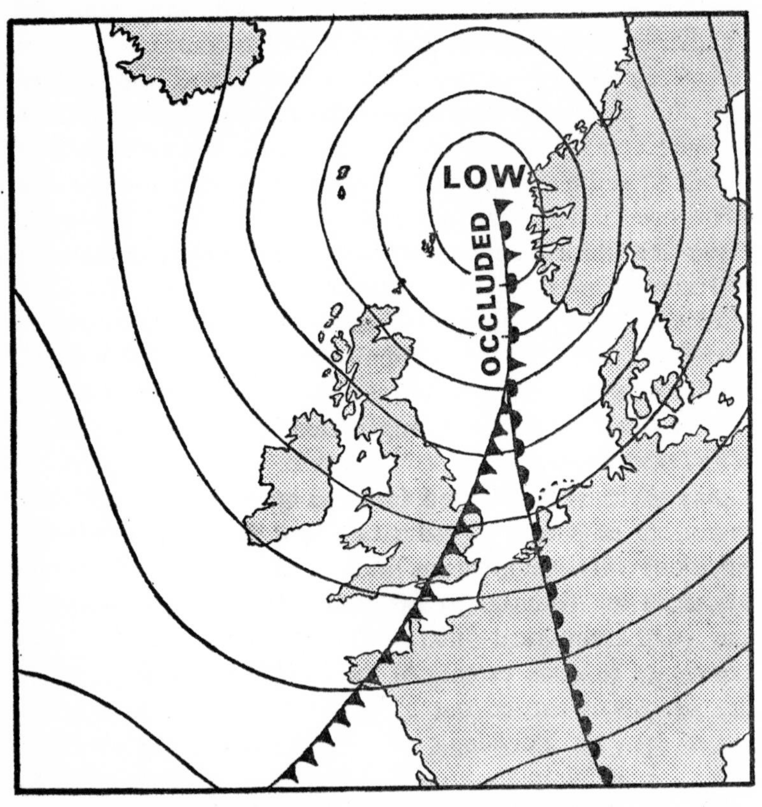

Fig. 7. Map showing fronts of a depression.

methods in forecasting to an absolute minimum is an objective that has the utmost priority, and it is being met in a number of ways.

In the first place research—much of it fully automated—is being carried out all the time by scientists attached to the Meteorological Office and by weather services abroad: to quote just one example, a project has recently been launched in the Scilly Isles to make an intensive study of approaching rainbelts. Secondly, increasing importance is being attached to developments within the upper atmosphere, because these frequently have a direct bearing on changes taking place at lower levels. To obtain information from regions high above the earth's surface, specially equipped weather stations release into the atmosphere balloon-carried meteorological instruments—these are called 'radio sonde' ascents. The instruments weigh a mere 3 lb and are capable of transmitting radio messages, as they ascend, concerning barometric pressure, temperature, and humidity. Their signal consists of a musical note of varying pitch which is continuously charted by the ground stations.

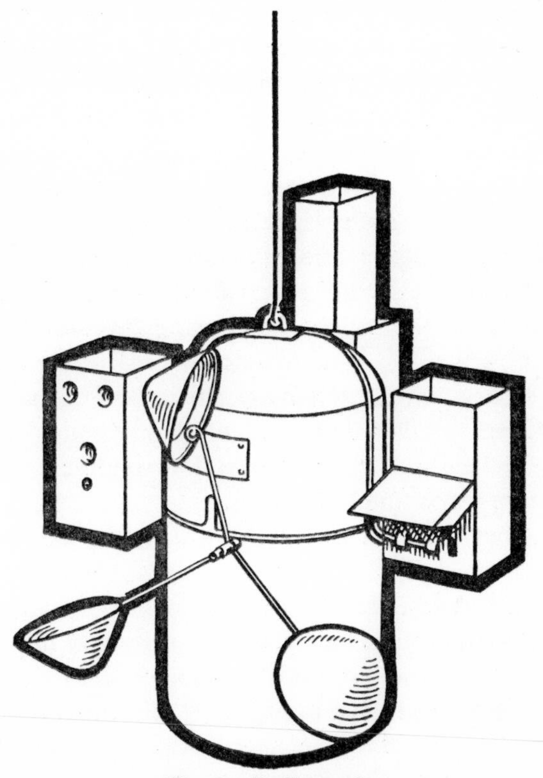

Fig. 8. Radio sonde.

Then, from a number of selected stations, reports are transmitted to the Central Forecasting Office concerning the strength and direction of upper atmosphere winds. These are again measured by balloons, which carry metallised reflectors and radar apparatus and are tuned to a master radar receiver at the ground stations. The invention, known as 'radar sonde', is a post-war one. Previously, hydrogen-filled balloons were tracked by theodolite to determine wind direction and strength at levels well above the ground, but once the balloon disappeared from sight the observations were cut short. Some of the latest balloons now reach heights of 100,000 ft before bursting.

During the last war and for some time afterwards regular meteorological flights were made by aircraft which flew at predetermined heights and radioed reports at regular half-hour intervals. The flights have now been discontinued, partly on cost grounds but also because the upper atmosphere charts constructed from balloon ascents are more comprehensive in many respects. But aircraft are still used for special research projects, and at South Uist, in the Hebrides, the upper atmosphere is regularly studied by means of rocket ascents.

On top of all this, the Met Office receives regular cloud photographs of the northern hemisphere from satellite flights. These are particularly valuable, as they enable the forecasters to keep a check on the relative positions and intensities of all the principal weather systems, but they are not sufficiently detailed to be used on their own for countries of variable weather that are as small as Britain.

In another field of research, which has an immediate application in everyday forecasting, radar devices are being widely used today by the Meteorological Office to locate thunderstorms, areas of hail, and other regions of heavy shower or general rainfall activity. It is possible to observe rain clouds on a radar screen within a radius of 70 miles, and this provides useful guidance to the forecasters when they are timing the arrival of rainbelts that appear on the plotted charts to be moving towards Britain. Radar apparatus also serves to give warning of any additional rainfall that has not been detected on the current charts. A further new technique, using radar, has been evolved to enable the meteorologists to measure rainfall without employing rain gauges, and an ambitious numerical technique, which is computer-aided, has been put into operation for providing estimates of rainfall amounts.

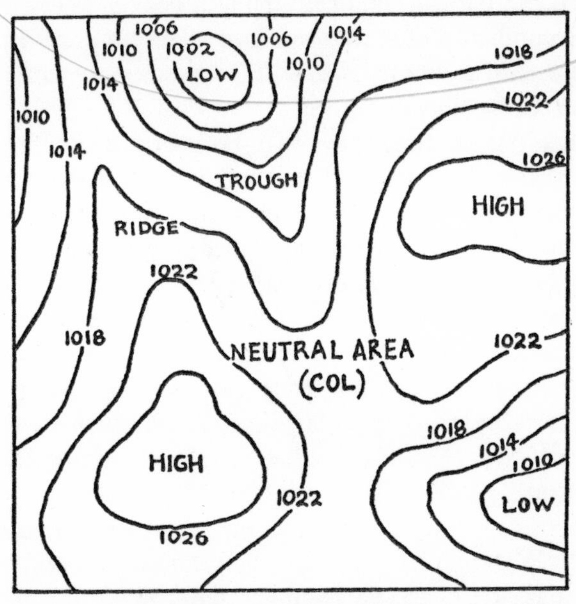

Fig. 9. Main pressure systems that create weather, showing a marked ridge and trough.

Meteorologists have invented a special nomenclature for different weather systems, as shown by the patterns by the isobars. The most important are the 'depression' and the 'anticyclone', which are the names given to regions of low and high barometric pressure respectively. The centres of these systems on the weather charts are accordingly marked with either an 'L' or an 'H'.

Elongated extensions from a depression, equivalent to long valleys on a relief map, are called 'troughs of low pressure';

they often occur at the frontal boundaries between conflicting airstreams, and here the weather is particularly unsettled and stormy. Similarly shaped extensions from an anticyclone, equivalent to hill ridges or escarpments on a relief map, are known as 'ridges of high pressure'.

How the different weather systems behave around Britain makes a fascinating study. The atmosphere here might well be regarded as a battleground between weather forces in a constant state of conflict, and there are times when no system prevails for more than a day or two, so that we then experience only 'samples' of weather.

Depressions are associated with belts of precipitation, often accompanied by strong wind, and anticyclones with dry weather and light winds near their centres but with rather stronger winds on their fringes. Sometimes an anticyclone will approach Britain but will slow down when the centre is still some hundreds of miles away, so that the whole of the country remains near the boundary between this system and the nearest adjacent depression. This kind of situation is known to experienced meteorologists as they watch their charts; they know, too, the favourite lines of approach of the high- and low-pressure systems.

Most of the large depressions that approach Britain begin their life in the North Atlantic a few hundred miles from the American coast in the approximate latitude 40–45° N. This is the area of the northern hemisphere where tropical and polar airstreams—and the ocean currents associated with them—are most frequently in conflict, and, at intervals, a comparatively small vortex of air will be created along the boundary line between the two main air masses that sends cloud-producing air spiralling upwards to a considerable height.

The depression, once formed, will be carried towards Europe by the prevailing winds, which are normally south-westerly. The speed of the system will depend upon the strength of this prevailing wind—some move as much as 600–700 miles a day—and during their early stages they may grow very large. By

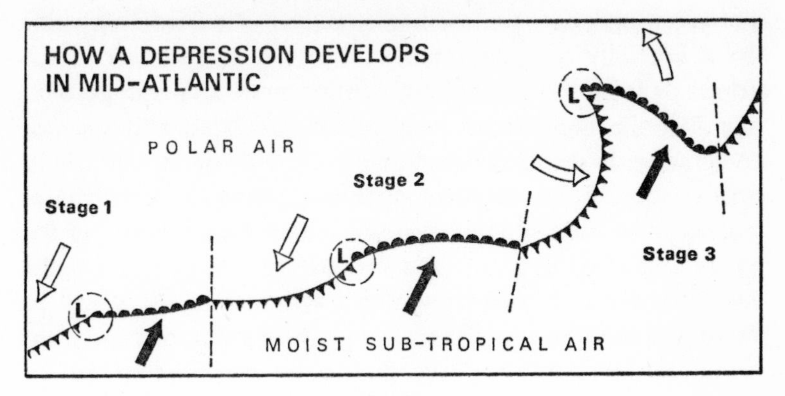

Fig. 10.

the time they are a week old they often cover areas of several thousands of square miles, but they normally slow down and weaken before reaching their destination, which is usually the Arctic—though a few reach the Mediterranean when prevailing winds in the eastern North Atlantic turn north-westerly.

The speed and strength of depressions as they approach western Europe varies considerably. Very active systems are described as 'intense' or 'vigorous'. The adjectives 'deep' or 'shallow' refer to the difference, either great or small, between barometric pressure at the centre and at the edge of the respective systems. Thus, on the weather charts, the deep depressions will be represented by many circles or oblong loops of isobars, and near the centres of the systems the isobars are likely to be close together, showing that winds are strong or blowing at gale force. Shallow depressions, represented by only a few isobaric circles or loops, will be less windy, and they normally form in areas where prevailing winds are slack, so that they move only slowly. But, like the deeper depressions, they produce rainfall; and the rainfall is likely to be heavy and thundery at times, particularly during the summer.

For forecasting purposes the size, speed and intensity of an approaching depression will be noted by the meteorologists as they study their charts. The direction it takes during any one period, whether it is a few hours, a day, or a week, may also be a crucially important factor. But the preparation of the forecasts from the charted information must take into account the fact the depressions carry their own wind and weather systems with them. Rainfall, with or without strong winds, will normally be strongly featured, likewise temperature changes, for each of these systems represents a mixture of the polar and tropical airstreams that are drawn into it. On the southern flank of a depression the predominant winds will be from the south-west, bringing muggy tropical maritime air, while, on the northern flank, cool polar maritime air will be carried southwards. Each airstream remains in continual conflict with the other in the depression's frontal boundary regions.

Due to its great size, it is not necessary for the actual centre of a large depression to cross Britain in order to give the country a wet and windy day. Even if the centre passes well to the north of Scotland, much of Britain—particularly northern districts—will have rain and mainly south-west winds from the depression's southern flank. But if the centre takes a more southerly track from west to east then the British Isles will have greater amounts of rainfall, as well as stronger winds. These will be south-westerly at first, but colder, from the west or north-west, in the rear of the system. If the depression approaches from due south, and the centre crosses the English Channel, then rainfall will be very prolonged. This type of weather is common during very unsettled and thundery periods in summer.

At other times depression centres will travel from Greenland or Iceland towards the Mediterranean, which will mean that the prevailing winds over Britain will be north-westerly and the weather here will be dominated by showery polar maritime airstreams and strong gusts or squalls of wind as each shower—or group of showers—passes overhead.

A feature of many Atlantic depressions is their ability to create secondary centres on their southern flanks. This happens very often when the main centre slows down in mid-ocean and when the frontal boundary at the southern extremity is almost stationary. As soon as they are formed, the secondaries detach themselves from the parent system and generally move rapidly north-eastwards, many of the centres actually crossing the British Isles.

It is these secondary depressions that normally give us our worst gales. Sometimes winds reach hurricane force, and as many as four or five secondary depressions belonging to the same family may reach our coasts during the course of roughly a week or ten days or so of very unsettled weather. But the winds of even the most violent storms affecting Britain are less strong and damaging than those of most tropical low-pressure systems, which are known in the West Indies and off the East Coast of North America simply as 'hurricanes', in the western Pacific as 'typhoons', in the Bay of Bengal as 'cyclones', and off the north-west coast of Australia as 'willy-willies'.

One quality common to all cyclonic disturbances in the northern hemisphere is that winds blow anticlockwise round their low pressure centres; but they do not blow parallel to the isobars. Instead, they are deflected towards the low-pressure centres at an angle of approximately 20 degrees to the isobars, as the weather charts indicate. This deflection is caused by the effect of the earth spinning on its axis. Unlike the depressions of middle and high latitudes, the intense tropical storms tend to move slowly and irregularly, like fast-spinning tops—which is what, in effect, they are.

Each storm that is created contains more energy than many nuclear bombs, but still more powerful are the anticyclones, the high-pressure regions that have completely the reverse weather influence. A strong or developing anticyclone or an extending high-pressure ridge from an anticyclone will deflect an approaching depression away from its boundaries, but this does not apply to a temporary high-pressure ridge that squeezes its way between two depressions for a short period before collapsing like a pricked balloon.

It is therefore necessary for the forecaster to estimate at

regular intervals, the strength, durability, and development power of all high-pressure regions on the weather charts. Barometric pressure readings on the fringes of each one must be constantly watched. If they rise simultaneously on one side of an anticyclone, the system will extend in that direction; conversely, if they fall simultaneously on one side of the system, the anticyclone will be weakening or retreating and may allow a low-pressure area to move in to take its place.

The movement of anticyclones affects wind and weather conditions in the opposite way from that of depressions. Winds blow clockwise round these systems in the northern hemisphere, and, as the charts show, they blow slightly outwards from the isobars as a result of the earth's spinning movement. But the strength of the wind becomes progressively weaker as one approaches the centre of a high-pressure system.

Anticyclones are regions of predominantly dry conditions, with air descending from upper levels of the atmosphere to the earth's surface. This action prevents the build-up of storm clouds, but anticyclones should not be regarded as inevitable bringers of fine weather. Much depends on the origin of each system, and this can vary from sub-tropical to polar.

The most persistent fine-weather producer in north-west Europe is the semi-permanent Azores anticyclone, which owes its existence to the fact that this region is one of the chief zones of descending air in the world's atmospheric motion. At periodic intervals—but less frequently in winter than in summer—it sends out an extending northward arm in the form of a high-pressure ridge which, if it is strong enough, will cross Britain towards southern and central Scandinavia and will give generally settled conditions along the entire route. In spring, summer, and early autumn this ridge will be synonymous with sunny, generally warm weather; but during late autumn and winter the weather will tend to be cloudy and rather dull, mainly dry in inland regions but with a little drizzle at times near coasts. Under the thin cloud layer that forms at such times in persistent high-pressure conditions, haze will collect beneath the clouds, making them look surprisingly stormy, though the nearest real storm clouds may be as much as 500 miles distant. Then, all too readily, the haze may thicken into fog, which gradually accumulates and spreads over many hundreds of square miles of countryside if there are no hills or mountain ranges to contain it.

Very dry summers are generally caused by the unusual persistence of the Azores anticyclone, but in unsettled summer seasons Atlantic depressions are the main weather features, while high-pressure ridges from the Azores anticyclone, though they may occur at times, are short-lived and comparatively weak. The forecasters, anxious to please the holiday public, as well as farmers with harvesting problems, hope vainly for the chance to give news of a fine-weather spell as each new weather chart is produced, but it may be a week or more before they find cause for very much optimism.

A less permanent anticyclone than the Azores system is the one that forms over Asia and eastern Europe during the winter. The seasonal cooling of this vast land mass chills the atmosphere above it to quite considerable heights and this creates a general downward motion of air, for cold air is heavier than warm air and always gravitates to the lowest levels.

This anticyclone has either a very profound or surprisingly little influence during any single winter season in Britain. When winters here are comparatively mild, frequent depressions bring south-westerly winds from the Atlantic; these travel across the North Sea as far as the Baltic, which is also the approximate limit of the anticyclonic influence from the east at this time. But during severe winters in Britain, and occasionally during warmer seasons, the Asian anticyclone becomes more than normally active and sends a powerful ridge of high pressure towards the North Sea, and this may eventually extend beyond the coast of Ireland in one direction and to the Mediterranean in another. The result is that the whole of Europe then receives an easterly airflow, with continental-type weather conditions. The winds are dry but precipitation belts, frequently taking the form of snow, form in the constantly moving boundary regions between the high-pressure influence to the east and the Atlantic low-pressure influence to the west. In summer, easterly winds are less frequent but when they occur they bring very warm weather, for at this season the Continent is then being strongly heated.

Early and mid-spring weather charts frequently show an anticyclone over the Atlantic, though this is a less persistent feature than the other two types. All too often it is a cause of unsettled, cold, and showery weather in this country, because its centre may move no nearer to Britain than a point approximately 600 miles west of Ireland, and as a result we then receive a strong northerly wind flow which brings in polar air and squally showers. But if the anticyclonic centre moves sufficiently far to the east, the weather gradually improves.

We are of course very used to the fact that, every year, there are times when our British weather gets into a 'rut'. The cause of this will normally be a marked persistence of either the anticyclonic influence or of the reverse. Once a powerful high- or low-pressure system develops in such a way as to influence large areas of country, the day-to-day conditions may be very similar for weeks on end, and for this reason one can say that up to a period of several weeks at least, the longer a fine or a wet spell has lasted, the longer it is likely to continue. The weather charts prove that this is so.

5

THE FORECASTS

THE standard short-range Met Office forecasts issued to the Press and to various radio and television channels for use by the general public cover a period of 12 hours. A further outlook is normally incorporated, which covers a second 12-hour period in very general terms. However, when the weather is settled and the atmosphere is stable—that is, incapable of producing rainfall—the period covered by the forecast may be extended to 72 hours altogether.

The main feature of the standard forecast is the amount of detail that it normally gives. It will mention the weather expected in terms of temperature, sunshine, wind, precipitation, and visibility, and if any change is expected during the period of the forecast an attempt will be made to give some idea of the time that this may occur.

Sometimes the forecasters are accused of phrasing their predictions in such a way as to cover all eventualities. A typical example of an all-embracing forecast runs something like this:

> There will be rain, followed by sunny periods with occasional showers at times. Winds will be mainly rather strong but will moderate and become light inland by evening. Although temperatures will be mainly below average, they will reach average or above average levels locally during sunny periods . . .

Surprising as it may at first seem, this forecast is a reasonably precise statement of what is expected. The rain referred to in the first part of the forecast is associated with overcast conditions, and, after the clearance of the overcast, sunny periods and showers are forecast alternately, which is quite a common occurrence. The fact that winds will moderate towards evening in inland regions is a normal feature of most showery days, and it is also a feature of this type of weather that temperatures vary considerably from place to place and are usually considerably lower in exposed windward regions than in sheltered valleys. This makes it impossible to make a standard pronouncement about temperatures that will apply to all districts.

It would sound better, no doubt, if the forecasters could give straightforward opinions about the conditions they expect in terms of whether it is likely to be warm, wet, cold, or dry. But such is the variety of weather that can occur during a single day that qualifying statements of various kinds are often necessary. It is important therefore that members of the public should understand how the meteorologists express their opinions, and because many English words describing the weather are used loosely in everyday speech it is necessary for them to define certain words more precisely: for example, in ordinary speech the word 'fine' can mean anything from brilliantly sunny to dry but with an overcast sky. In the forecasts the word is used for occasions when the weather is dry and free from fog, with some sunshine during the daylight hours. But when the weather is expected to be dry but with persistent low cloud and no bright periods, the word 'dull' is used. Similarly, the words 'bright', 'bright periods', 'bright intervals', and 'cloudy' all have reasonably precise meanings in the forecasters' vocabulary (see Appendix 8).

In many forecasts it is impossible to exclude some element of uncertainty. This tends to increase with the period of future time that a particular forecast covers, but the nature and extent of the uncertainty will vary according to circumstances. On one occasion it may lie in the timing of a weather event and on another with the type of precipitation expected. The varying degrees of confidence or uncertainty are expressed in the following way:

1. The first or highest degree of certainty is expressed by the omission of qualifying words or phrases: eg, 'fine today', 'rain tomorrow', 'winds will be westerly'.

Fig. 11. Areas used in the Meteorological Office regional weather forecasts broadcast by the BBC.

2. The second degree of certainty is indicated by the terms 'expected', 'probable' or 'probably', 'likely', or 'likelihood of'.
3. The third degree of certainty is indicated by the terms 'prospect of', 'indications of', 'conditions are favourable for'.
4. The fourth degree of certainty is indicated by the terms

'may be', 'may occur', 'chance of', 'perhaps', 'possible', or 'possibility of'.

Other qualifying words are used with their natural meanings, for example, 'mainly', 'temporarily', and 'occasionally'. They are necessary to enable the forecaster to express as clearly, unambiguously, and completely, as possible the knowledge he has of the future weather.

A definite commitment in a forecast without any qualifying phrase is always more likely to inspire confidence than an indefinite one, but, argues the Meteorological Office, if there is an element of doubt as to whether rain or snow, or whatever else is indicated, will occur, there is no point in giving a verdict one way or the other merely for the sake of being definite.

The adjective 'official' is sometimes used by BBC announcers in connection with a current 12-hour forecast. This simply means that it has been issued by the Met Office, which is the basic source of all the short-range forecasts in the country; but the choice of words is perhaps unfortunate in that some people get the impression that the contents of the forecast are guaranteed to be accurate. And this is certainly not the intention.

Despite the increasing influence of television, the most widely used Met Office forecasts are those issued by BBC radio. Over the years these have increased considerably in number, and various improvements have been made in their general presentation. Every now and again changes are made in the times that the forecasts are issued, so it is advisable to check with the *Radio Times*. Radio 4 provides national and regional forecasts before the main news bulletins, each of the latter being concluded with a summary of the latest forecast for the benefit of listeners who have just tuned in. The weather/news sequence is thus: (1) national weather forecast, beginning with a general description of the movement of weather systems affecting Britain; (2) regional weather forecast; (3) national news bulletin; (4) summary of national forecast; (5) regional news bulletin;

Page 71: *(above)* Forked lightning at night, showing feeder strokes; *(below)* giant hail (with golf ball in centre) photographed at Melksham, Wiltshire, immediately after fall, 13 July 1967.

Page 72: *(above)* Hill fog (*stratus* cloud); *(left)* fair-weather coastal haze, near Bude, Cornwall; *(below)* BBC television mast at Crystal Palace protruding above fog layer (*stratus* cloud covering ground).

(6) summary of regional weather forecast. Transmissions are concluded at approximately 11.50 pm with a late-night coastal waters forecast.

A regular schedule of weather forecasts is also given on Radios 1 and 2 after the news bulletins, the times of the transmissions being given in the *Radio Times*. The shipping forecasts, which cover all the sea areas surrounding the British coasts, are given on Radio 2 only; these also incorporate coastal waters forecasts. Radio 3 transmissions, like those of Radios 1 and 2, provide weather forecasts with each news bulletin.

Extreme economy of words is used in the shipping and coastal waters forecasts in order to present as much information as possible within a short space of time, so those who are interested in following them might find it necessary to have paper and pencil handy well in advance (see pp 234 and 235). After a brief general summary of the current weather situation, the forecast is given for each sea area in terms of wind strength and direction, general weather, and visibility. In the case of the coastal waters forecasts that follow, the presentation is similar, but barometric pressures are also given and, finally, the barometric tendency for each station, ie, whether the barometer has been rising or falling.

Gale warnings are issued immediately they are received; programmes now being specially interrupted for this purpose, though some years ago they were only issued at convenient programme junctions. The warnings are also stated, if in operation, at the opening of each shipping forecast or coastal water forecast when the latter is issued separately. Apart from gale warnings, special 'flash' weather messages—another innovation of recent years—are issued by interrupting programmes for a few seconds when any expected weather conditions are likely to be dangerous to a large section of the public. Flood, fog, ice, and snow warnings are covered by this special service, and the warnings are also repeated in the next news bulletins.

The Met Office forecasting service for the Press is almost as

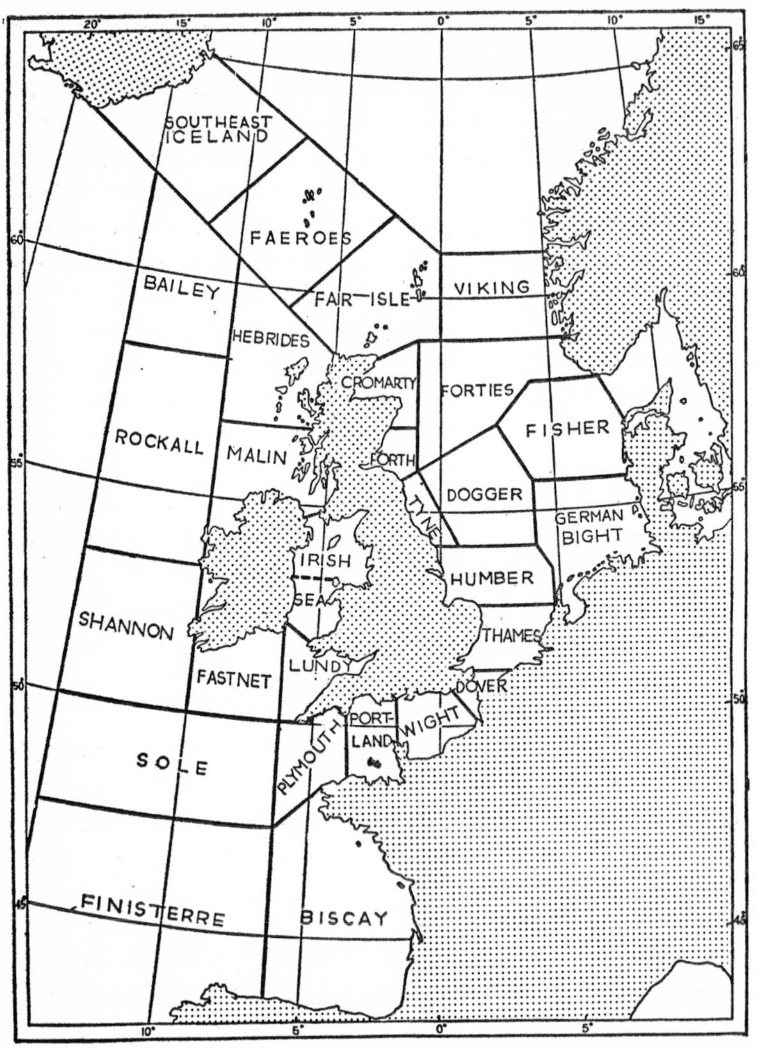

Fig. 12. Sea areas used in the Meteorological Office shipping forecasts and gale warnings broadcast by the BBC.

comprehensive as that given to the BBC, but there is an inevitable delay of up to 6 hours between the issue of the forecasts and their subsequent appearance. It is therefore advisable to check newspaper forecasts with the latest weather bulletins on radio. Among newspapers that publish detailed forecasts for the various regions of the British Isles are *The Daily Telegraph*, *The Times*, *The Guardian*, and *The Scotsman*.

In addition, some newspapers also publish charts showing actual weather conditions in the northern hemisphere on the day before publication, together with smaller charts shewing expected future weather conditions. Some of the popular daily newspapers publish forecasts that are too sketchy to be of very much use, and the accompanying diagrams give artists' impressions of what is expected, not the basic chart issued by the forecasters.

The type of forecasting service available on BBC and ITA television channels combines the pictorial advantages of the better newspaper forecasts with the up-to-dateness of the radio bulletins. What is more, most of the television weather bulletins on the BBC channels are given by a Met Office forecaster, who is able, when time permits, to say a few words about the current situation. In any case, the latest charts, both actual and predicted, are shown. The late night BBC television forecast can be very helpful during the summer to farmers, yachtsmen, climbers, and holidaymakers in general, so long as they remember to make a check on the situation during the following morning.

The main disadvantage of the BBC television forecasts is the fact that none appear before the afternoon, and the majority of viewers may see only one televised forecast during the course of a day.

The television services of both BBC and ITA take part in the flash weather message scheme whenever an imminent threat of weather is likely to cause serious inconvenience to a large number of viewers. The main ITA forecast bulletins are run on similar lines of those of the BBC, but the information

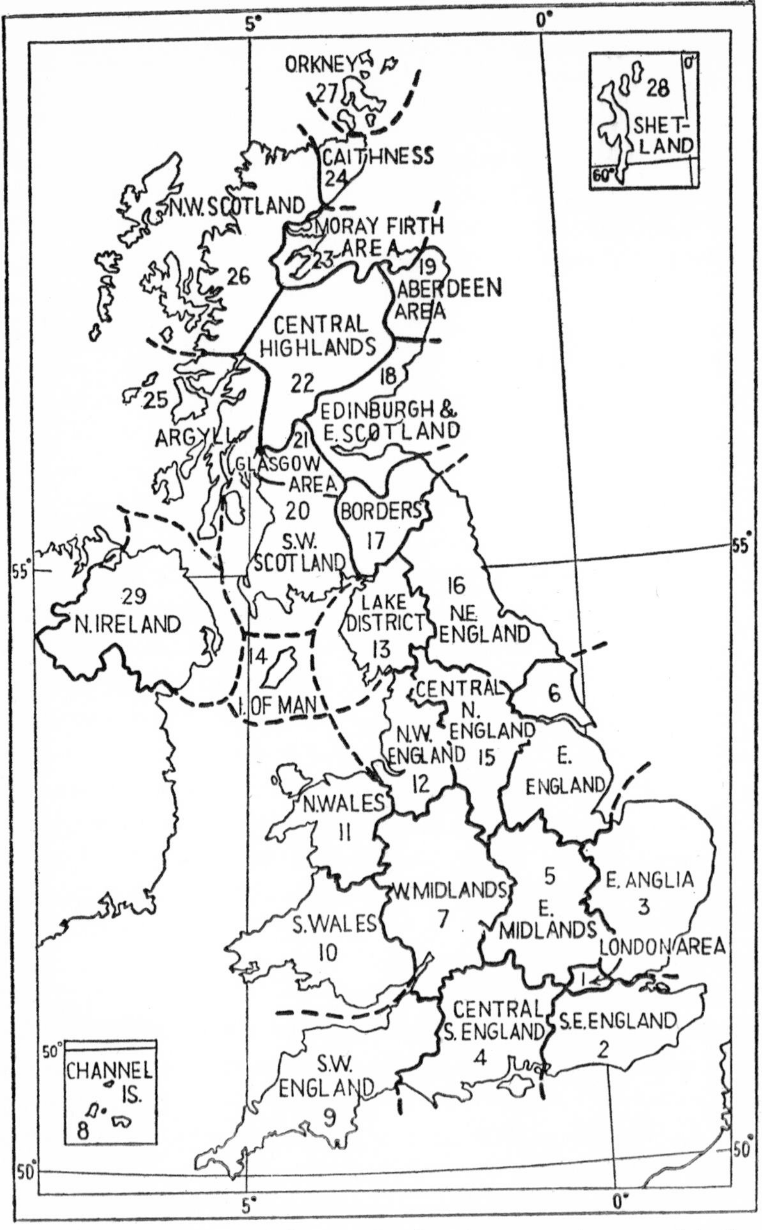

Fig. 13. Land areas and the counties that these comprise as used in the daily Meteorological Office forecasts published in the press.

1. *London Area*
County of London; Middlesex.
2. *Southeast England*
Kent; Surrey; Sussex.
3. *East Anglia*
Norfolk; Suffolk; Cambridgeshire; Essex.
4. *Central Southern England*
Berkshire; Hampshire; Wiltshire; Dorset; Isle of Wight.
5. *East Midlands*
Leicestershire; Rutland; Northamptonshire; Huntingdonshire; Bedfordshire; Hertfordshire; Oxfordshire; Buckinghamshire.
6. *East England*
Lincolnshire; Nottinghamshire; Yorkshire (East Riding).
7. *West Midlands*
Staffordshire; Shropshire; Worcestershire; Warwickshire; Herefordshire; Gloucestershire.
8. *Channel Islands*
9. *Southwest England*
Somerset; Devonshire; Cornwall.
10. *South Wales*
Cardiganshire; Radnorshire; Pembrokeshire; Carmarthenshire; Breconshire; Glamorgan; Monmouthshire.
11. *North Wales*
Anglesey; Carnarvonshire; Denbighshire; Flintshire; Merionethshire; Montgomeryshire.
12. *Northwest England*
Lancashire; Cheshire.
13. *Lake District*
Cumberland; Westmorland.
14. *Isle of Man*
15. *Central Northern England*
Yorkshire (West Riding); Derbyshire.
16. *Northeast England*
Northumberland; Durham; Yorkshire (North Riding).
17. *Borders*
The counties of Berwick, Roxburgh, Peebles, Selkirk and that part of Dumfries-shire east of Annandale.
18. *Edinburgh & East Scotland*
The part of Kincardine southward of a line running west from Stonehaven.
Angus southeast of a line Edzell to Alyth.
Perthshire south and east of a line through Alyth, Dunkeld, Methven and due south from Methven to the borders of Kinross. Fife, Kinross-shire and Clackmannanshire, East Stirlingshire (East of the Campsie Fells), West Lothian, Midlothian, and East Lothian.
19. *Aberdeen Area*
The whole of Aberdeenshire except that part lying west of a line from Mount Battock to the Buck.
The part of Kincardine northward of a line running west from Stonehaven.
20. *Southwest Scotland*
Dumfries-shire west of and including Annandale, Kirkcudbrightshire, Wigtownshire and Ayrshire.
Lanarkshire except that part of the Glasgow area as defined below. Bute and Arran.
21. *Glasgow Area*
The whole of Renfrewshire.
The industrial part of Lanarkshire which includes Glasgow, Airdrie, Motherwell, Wishaw, Carluke, Lanark, Hamilton and East Kilbride.
The county of Stirling southwest of the Campsie Fells including Kilsyth and Drymen.
That part of Dunbartonshire which includes Dumbarton and Alexandria and all places southwest of a line through Dumbarton, Alexandria and Drymen.
N.B.—There is part of Dunbartonshire between Lanarkshire and Stirlingshire. This is also included in the Glasgow area.
22. *Central Highlands*
Parts of Inverness-shire comprising the Great Glen and the area east of the Great Glen (except the area within 10 miles of the Burgh of Inverness).
Aberdeenshire west of a line from Mount Battock to the Buck.
The highland parts of Nairn, Moray and Banff.
Angus northwest of Strathmore.
Perthshire north and west of a line through Alyth, Dunkeld, Methven and due south from Methven to the boundary of Kinross, including Alyth, Dunkeld and Methven and thence along the Ochils.
Parts of the counties of Dunbartonshire and Stirling north of a line through Dumbarton, Alexandria, Drymen, the Campsie Fells and Stirling, but excluding these places.
23. *Moray Firth Area*
Bonar Bridge and the eastern coastal strip of Sutherland. (This does not include Lairg.)
Ross and Cromarty east of Ben Wyvis and Ben Tharsuinn. The Burgh of Inverness and the county of Inverness within 10 miles of the county town.
The lowlands of the counties of Nairn, Moray and Banff.
24. *Caithness*
25. *Argyll*
The county of Argyll.
26. *Northwest Scotland*
The whole of Sutherland except Bonar Bridge and eastern coastal district (Lairg is included in this district).
The whole of Ross except that part east of Ben Wyvis and Ben Tharsuinn.
Inverness-shire west of the Great Glen (except the area within 10 miles of the Burgh of Inverness). The island parts of Inverness-shire and Ross.
27. *Orkney*
28. *Shetland*
29. *Northern Ireland*
Londonderry. Antrim; Tyrone; Fermanagh; Armagh; Down.

is normally presented by members of their own staffs, based on data and captioned display charts provided by the Meteorological Office. But the presentation does not normally include charts, as used by the BBC, that show up the high- and low-pressure systems and the main frontal regions of depressions, together with the general pattern of the isobars.

One thing is particularly apparent these days—both radio and television are becoming more forecast-conscious. This applies especially to sporting and other outdoor events of national or regional importance, and, in all the forecast bulletins, local weather differences within a broad general region are mentioned whenever possible. The introduction of local radio stations has now taken this ideal a further step forward, for each new station to be opened has negotiated with the Met Office for the provision of detailed forecasts for its particular area.

In addition to providing the general public with forecasts through the mass media, the Met Office is willing to provide individual forecasts for specific purposes. No charge is made for answering a query if this can be done instantly, as is often the case. But if a special briefing or schedule of forecasts is required, then a charge is made according to the amount of work involved.

Individual forecast queries are handled by the appropriate regional offices or weather centres. Basically, a weather centre is an offshoot of a regional meteorological office that is situated in an accessible position for most members of the public, and specially equipped to deal with enquiries from personal callers and by telephone.

Most of the weather centres have large windows displaying current information about actual and forecast weather conditions for all parts of the country. In addition to answering general queries from members of the public, the centres provide routine forecasts for public organisations such as the gas and electricity boards, which are particularly interested in adverse winter conditions such as low day and night temperatures, snowfall, and strong east winds. These conditions cause abnormally heavy loads, and preparations to meet these must be made in advance whenever possible.

The variety of purposes for which forecasts are required is very great, and is in fact increasing every year. River boards receive regular warnings of heavy rainfall likely to lead to flooding, and county and local government authorities, who are also regular Met Office customers, need to know about the likelihood of abnormal conditions such as floods, snowfall, and icy roads. The Ministry of Transport requires frequent forecasts during the winter months, and takes into account the particular vulnerability of motorways to fogs and high winds.

Then there are many industrial and commercial firms who apply to the weather centres for particular forecasts. For example, many food industries are affected by temperature and sunshine, particularly ice cream. The large ice-cream companies, with their constantly expanding cold store and transport facilities, are generally less vulnerable nowadays to sudden heatwaves or the collapse of fine weather than the smaller operators who manufacture from day to day. By contrast, cold weather greatly accelerates the demand for bread and cakes, and manufacturers and retailers require constant advice on this subject. The building and civil engineering industries, however, are probably the most weather-sensitive of all. Sites and partly completed buildings can be flooded or frozen, and high structures adversely affected by strong winds or gusts.

The Met Office has recently made special studies of the requirements of the food and building industries, just as it did some years ago for the needs of farmers—another big group of people who receive constant advice from the weather centres. Such are the proved connections between weather and what happens on the land, farmers can be warned in advance of conditions likely to lead to sugar beet virus, potato blight, liver fluke in sheep, and gastro-enteritis in calves.

The London Weather Centre, which is in Kingsway and was the first to be opened, has now created an interesting precedent by incorporating a sound broadcasting studio, so that the regional South-east England weather forecasts of the BBC can be transmitted live from here. This centre handled over 114,000 enquiries during 1959, its first year of operation, and the number has now more than doubled. A second weather centre was opened in Glasgow during 1959, another in Manchester during the following year, and a fourth in Southampton in 1961. Then came a gap before the opening of the Newcastle centre in April 1967. In the Midlands the absence of a weather centre has been keenly felt for many years. The regional meteorological office is at Watnall, Nottingham, and receives its night forecasts from the London Weather Centre, and another regional office is situated at Birmingham airport. An analysis of the various types of enquiries handled by the weather centres during their first seven years of operation showed that at the London and Glasgow centres the greatest number came from holidaymakers, that transport enquiries were the most numerous at the Manchester centre, while marine enquiries took the first place at the Southampton centre.

Since 1960 local area weather forecasts have been prepared by the Meteorological Office and recorded by the GPO. The recordings are changed, on average, four times daily but more frequently than this if the situation demands it. Like the weather centre idea, the scheme began in London where it was known as the WEA telephone service. Here, during the first year, nearly 2,500,000 WEA telephone calls were made, and the average annual growth rate since then has been about 10 per cent. Over 4,000,000 calls, however, were made in 1963 as a direct result of the severe winter, compared to 3,000,000 in 1962 and 3,500,000 in 1964.

Area directories give details of the automatic telephone weather services available now in many towns, and, as when the service began, there is no charge to the public apart from the cost of the call. For foreign tourists in the London area recorded local weather forecasts are available in French, German and Spanish.

Nearly one-third of the Met Office staff is employed in providing a world-wide forecasting service for the Royal Air Force. Separate meteorological offices for this purpose are located at ninety stations that stretch from the UK to Germany, the Mediterranean, and the Middle East, thence to the Far East as far as Borneo.

Civil aviation forecasts in Britain are dealt with by airport meteorological offices. Where these are situated in areas not conveniently served by regional met offices or weather centres, the forecasters here also provide information for the general public, industry, and local authorities. In the UK the provision of forecasts for medium- and long-range flights operating above 5,000 ft is the responsibility of the Principal Forecasting Officer at London (Heathrow) Airport, and since October 1967 Heathrow has been the European forecasting centre for aviation. It can supply forecasts suitable for direct use for flights over a very wide area, including the North Atlantic, North America, and much of the polar region. The office is specially equipped to receive satellite cloud pictures, in addition to a constant flow of information and facsimile charts from the Central Forecasting Office at Bracknell. The special short-range forecasts of upper winds and temperatures up to a height of 40,000 ft are now largely based on computed data provided by the main Met Office computer.

In 1967 an experimental trial of a new service—the weather-routing of ships across the Atlantic—was carried out by the Met Office Marine Branch. Four ships took part and agreed to follow courses selected by the meteorologists to give the quickest crossing and minimum buffeting from heavy seas. One ship gained 14 hours on its normal route, which represented a saving of fuel and possible damage to its cargo.

The service is now fully operational. Just before sailing the master of a ship contacts the Met Office at Bracknell, which gives advice covering at least the next 48 hours. This is very important. For example, should a ship leaving Liverpool go north around Ireland or will it pay to go the longer way via Fastnet? The shortest route is not necessarily the quickest.

The initial briefing is followed by further advisory messages transmitted to the ship by Portishead Radio every 48 hours, but more frequently if desired. All the messages advise on the best course to follow and include, as and when required, special 48-hour weather forecasts incorporating estimates of wind and sea conditions.

Another service recently introduced by the Meteorological Office concerns the weather-forecasting arrangements for the ski resorts at Cairngorm, Glenshee, and Glencoe. The three ski-lift companies are co-operating. They telephone reports of existing weather over their respective areas to the Glasgow Weather Centre by 4 pm each day, and from here the information is transmitted to the Central Forecasting Office at Bracknell for use there and for redistribution nationally. Then, at 6 pm, the Glasgow Weather Centre issues all the weather reports to the Press Association, together with a special ski-ing prospects forecast for the next day. This covers the state of the sky, precipitation (if expected), freezing level, and wind, and on Friday evenings the forecast becomes an extended one and refers to the weather for the next two days. There is no doubt that the provision of this new service can enable skiers to plan a weekend at the resorts as late as Friday evening; they can travel by air at the last moment from as far away as the south of England.

The improvements since the war in the presentation of the weather forecasts and in the widening of their scope have tended to overshadow the attempts to improve their general standard of accuracy. The public, quite naturally, is particularly interested in this aspect and is sometimes disbelieving. But checks on the individual elements mentioned in the forecasts are constantly being made by the Met Office, who find that the 12-hour forecasts as a whole have an average accuracy of at least 70 per cent. Those that are issued for the day-time period during the previous evening give an accuracy of between 70–80 per cent, while those issued on the morning of the same day are between 80–85 per cent accurate. These figures apply generally to all the regions, coastal as well as inland.

A criticism of the national weather service—and one that is frequently made by mariners—is that the forecasts are spaced too widely and the whole complexion of a particular weather situation may alter considerably in the time between any two of them. The argument is particularly valid during times of emergency. For example, the midnight forecast on 14 January 1968 made no mention of strong winds for northern areas of Britain, yet, 6 hours later, winds of well over 100 mph stripped Clydeside of thousands of roofs and made many people homeless. In the same year, in July, Devon and Somerset experienced disastrous floods, but the majority of people were unprepared.

Another great flood affected South-east England on 15 September. River boards were sent special warnings on the previous day, but the public was totally unprepared for the catastrophe. And so it was with the North Sea flood of late January and early February 1953, when coastal dwellers were taken unawares, and with the one that destroyed most of Lynmouth, in North Devon during the previous August (see Appendix 3). Another time when the forecasters and the general public were taken by surprise occurred on V-day 1946, which was unexpectedly wet. Then, in 1947 the thaws of the spring did not materialise as soon as predicted—they turned out to be blizzards. And so, much later, did a forecast thaw during the second week of February 1963 when advancing warm air masses from the Atlantic, which were just beginning to be effective, were pushed back by easterly winds.

For those who had the time to listen to them regularly, the former dawn-to-dusk 'Airmet' weather broadcasts were very helpful at times like these. They were started before the last war and were intended at first for private fliers and small airfields which did not have a meteorological service. When they were restarted in 1946 they came from a small Met Office studio at Dunstable, and gave hourly reports of barometer pressures, precipitation, wind strength and direction, the amount of cloud, and the height of the cloud base. These were followed by informal district broadcasts in the style of the wartime weather briefings for aircrews.

Within a short space of time Airmet was rightly hailed as the most perfect weather-warning system ever devised, with the power to save lives, money, and general human frustration. Coastal fishermen used it, so did caterers, racing pigeon clubs, glider pilots, fruit growers, industrial concerns, and many other groups, including housewives.

To farmers it was a real boon. A Norfolk farmer reported:

> I always remember one harvest when rain was approaching and I had a field of shocked corn ready to cart. If rain was imminent it would be best left on the shock; but if there was time, it was fit to cart and stack. Airmet supplied the answer; that a belt of rain was moving northeast at 30 mph and that it had just commenced to rain at Hereford.
>
> A quick calculation indicated that it would reach us about 5 pm. So we set to work. At 4.55 pm we had the stack finished and covered up: at 5.10 pm the skies opened and there was no more corn carted for a week.

One of the great advantages of Airmet was the fact that it was broadcast on the long-wave transmission of 1,224 metres, and virtually anyone could pick it up with an ordinary radio receiver. But in March 1950 came the great blow, when the Copenhagen International Conference failed to allocate a wavelength for this service. Many protests have been made in Parliament since then, a monster petition was organised for its return, and suggestions have been made for using other broadcasting channels, including VHF. The official reply to questions was always that the service has been temporarily suspended, but there are some who take the view that the BBC weather bulletins, which have become more numerous, detailed, and regionalised since the closure, fulfil much the same function as Airmet and makes its return superfluous.

This view, however, is strongly challenged by those who used the service. Frequent as the present-day weather forecasts undoubtedly are, it cannot be claimed that they provide a continuous reporting service, and they offer less scope than Airmet did for correcting individual predictions that have gone wrong. Nor can they compete with Airmet for keeping the public in touch with the situation on days when weather movements are particularly difficult to interpret. This sometimes occurs on Bank Holidays, when there is a tendency for the forecasts to be over-cautious. The Whitsun periods of 1950 and 1956 are typical examples.

For these many reasons—even acknowledging the fact that short-range forecasting should become gradually more accurate in the course of time—it is to be hoped that Airmet will return without too much further delay. The demand for the service still exists; in fact it is potentially much greater, due to the increasing need for up-to-the-minute weather information in all sections of daily life. Mr Reginald Bennett expressed the position very aptly in the House of Commons some years ago when he was pleading for Airmet's restoration and criticising the lack of proper support for it from the Post Office. 'Our ministers,' he declared, 'are not yet prisoners of their past decisions.'

One encouraging piece of news recently announced by the Met Office Director-General is that, given adequate observations and computing facilities, it should be possible to produce reliable forecasts of the basic weather patterns for 5–7 days ahead. This is a considerable advance from the normal pattern of 12-hour forecasts and the occasional extended-range forecast when conditions permit. The optimism is due mainly to the very good progress being made in the use of computers, coupled with ever more thorough research into the physical processes taking place in the atmosphere. An increasing number of weather stations will, however, be needed to provide more observations, and the new fully automatic weather stations designed by the Meteorological Office will play an important role in this respect during the coming years.

In a different category entirely from the detailed short-range forecasts are the 30-day outlooks which have been provided by the Met Office at fortnightly intervals since December 1963. These are still regarded as experimental and a note to this effect is provided with each forecast, together with a statement that 'the confidence that can be placed upon the conclusions is less than for forecasts of weather for a particular locality for short periods ahead'. The outlooks can be supplied to individuals direct from the Meteorological Office, Bracknell, Berks at a small cost, but they are also released through the mass media.

Probably only very few people have kept regular notes on the accuracy of these forecasts, but after the first 4 years of issue the Met Office concluded that 73 per cent of the forecasts issued showed at least a 'moderate agreement' with the weather situations subsequently experienced. Precise accuracy ratings are difficult to obtain due to the fact that the forecasts cannot be given in other than rather vague general terms at the present time. Certainly only a few of the forecasts score what the Met Office describes as its highest rating, meaning 'no serious discrepancy'; but, similarly, only a few of the forecasts have qualified for the lowest rating—'no real resemblance' to the conditions subsequently experienced.

The meteorologists concerned with these long-range outlooks claim that if one eliminates rainfall, which is difficult to predict in inches the standard of accuracy improves. They also claim that they forecast temperatures correctly about two times out of three, that specific mention of snow or frost has been correct on roughly three out of four occasions, and fog or thunderstorm forecasts have been right on roughly eight out of ten occasions.

The method used to produce the 30-day outlooks is entirely different from the one used in short-range forecasting and cannot be used to time the arrival of any particular type of weather with precision. A 30-day outlook for, say, the month of April, which is prepared at the end of March, involves the meteorologists in the first place with a study of the daily weather charts for the whole of the March in question or for as much of this month as is available at the time, followed by a comparison with previous Marches that had similar characteristics. If, for example, March 1958 is the one that is found to be most similar to that of the present month, then it is thought reasonable to assume that the weather of April 1958 may to some extent be repeated during the month ahead. This analogue method is being constantly refined, and much of the necessary comparison work each time an outlook is prepared is being carried out by the Met Office computer.

The basic long-range forecasting technique was introduced to this country by the Americans, who used it with considerable effect during the last war, but it is a mistake to think that accurate and detailed long-range forecasts will come soon. We first need to find out more about the basic causes of weather changes and the way in which they operate.

In 1966 the Director-General of the Meteorological Office gave details of a survey that he had carried out to assess the role of weather science in the national economy. In a single year the Met Office expects to supply over 1,000,000 forecasts for aviation, including 400,000 special aviation briefings; it also answers over 1,000,000 enquiries from industry and the public, produces 25,000 weather bulletins on BBC sound and television channels, and deals with 10,000 requests for information on climate.

In the assessment of the economic value of meteorological services to the nation, the Director-General considered that the civil national weather service was worth £50,000,000 to £100,000,000 per annum, for a cost of £4,000,000, resulting in a probable overall benefit/cost ratio of about 20 to 1. So far as services for the general public were concerned, the estimate was that the daily advice was worth at least 2d per family, which amounted to £30,000,000 in a year or fifteen times the cost of the basic service for industry and the general public (see Appendix 5).

In certain situations the weather services required by specialised industries are not readily available from state organisations. The services provided in the Persian Gulf for oil companies are a case in point. Here an offshore weather-forecasting service was initiated by British Petroleum and Shell in 1957, and, at the time, weather advice for movements of the Shell exploration platform had to be based on the sketchy weather information available from a meteorological office some 100 miles away and from temporary observation points. However, the loss of a platform in 1956 brought home to those concerned the need for a specialised warning service. A number of new weather stations was eventually set up by the participating oil companies, and the technical work was co-ordinated by the International Meteorological Consultant Service (IMCOS). In recent years the system has been applied to other parts of the world, including an area of the Nigerian coast, the Adriatic, and the North Sea.

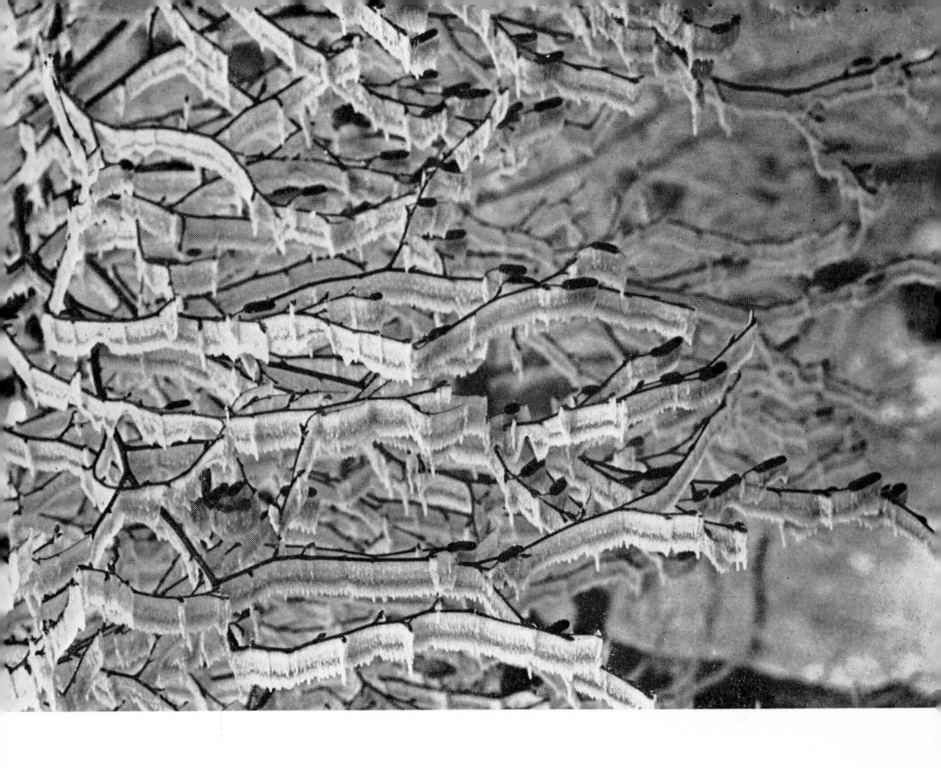

Page 89: *(above)* Heavy rime icing on trees; *(below left)* Whitstable, Kent in February 1956, when the sea froze; *(below right)* giant snowdrift resulting from a blizzard in February 1947, near Belah Viaduct, British Railways North Eastern Region.

Page 90: *(above)* Frost protection, using orchard 'smudge pots'; *(below left)* silver iodide generator used for cloud seeding; *(below right)* nylon-mesh screen used in the project at Gibraltar in which water is extracted from clouds passing over the Rock.

6

THE BRITISH CLIMATE

IN Britain we tend to be somewhat preoccupied by our constantly varying weather. Because of this it is easy to forget that the climate, as measured by the long-term distribution of temperature, rainfall, sunshine, and weather conditions generally, is itself always changing—but to a varying and irregular extent. Long-period fluctuations are imposed upon shorter ones, and, in a man's lifetime, the changes may be so small, compared with the ordinary day-to-day and year-to-year variations in weather conditions, that they are hardly noticed. Yet, in the course of time, the effects can be far reaching.

From an analysis of many kinds of evidence—based on variations in lake and river levels, the sequence of flora and fauna, and other historical records, including the instrumental readings of the last 300 years—we have a fair picture of our climate from 5000 BC to the present day, and, in particular, of the many remarkable changes that have taken place during the past 2,000 years. These are well worth careful study, for all the time the climatic pendulum swings first in one direction and then in the other.

About 500 BC, two hundred years of mainly dry warm weather came to an abrupt end. Lakes rose all over Europe, villages were destroyed by floods, and Alpine mining settlements were abandoned. The climate had improved in the south of the Continent but was still 'foggy, raw and damp' over Britain when Julius Caesar first invaded our shores in 55 BC. Through the action of unfavourable winds and tides he lost many of his transports on the beaches where Walmer and Deal have since been built. A gradual improvement took place after this, and when Agricola's successes in this country threw great lustre on the Roman arms at the end of the first century AD, the climate is thought to have been less humid than it is today.

By the beginning of the second century, a warm dry period had set in, which, according to some historians, lasted for 80 years with little interruption. Open-air theatres in Britain became scarcely less feasible than they were in Florence and Rome, though Roman attempts to cultivate the vine here met with only little success, possibly because of the poor quality of the soil. In the period AD 180–350 the climate deteriorated, but by the end of the Roman occupation it became warmer and drier again.

The great penetration of the Roman forces cannot be explained purely in terms of naval or military prowess, or by their constructional skill and their ability to fell trees, drain marshes, and heat their homes. One of the most crucial matters was the water supply, and here the Romans were very fortunate. Underground water tables stood very high, much higher than at present. For example, a Roman bucket was discovered at the bottom of a well in the Romano-British village of Woodyates, Dorset, which is more than 60 ft above the level to which modern wells have to be sunk in the immediate neighbourhood. Again, ancient Roman settlements have often been found near what are now very high springs, only usable in the wettest seasons. They were obviously perennial in Roman days, or nearly so, for it is hardly likely that the sites could have been chosen without adequate water supply.

But in the post-Roman era there were hard times, and during the Saxon period came the longest and most severe periods of drought that this country has experienced during the past 2,000 years.

Before the seventh century, drought had been infrequent but fierce. One in AD 484 was described as 'drying up all springs and rivers' and in 605 another drought was experienced 'with scorching heat'. By 676 there was a trend towards persistent outbreaks—they recurred annually for seven years; and another long series of droughts began in 713, no fewer than fifteen being recorded by the chroniclers between that year and 775. In these there are reports of 'a dry summer', a 'very hot summer', a 'burning drought', and a 'long and terrible drought with heat'.

The Saxons had no choice but to adjust their economy as best they could to meet the conditions, though they were helped by gaps between the greatest droughts when the weather was less severe. Nor were they denied an occasional picturesque incident. The Venerable Bede records, in his *Ecclesiastical History of England*:

> Bishop Wilfred, while preaching the Gospel to the people (the south Saxons, AD 681), not only delivered them from the misery of eternal damnation, but also from a terrible calamity of temporal death. For no rain had fallen in that district for three years before his arrival in the province, whereupon a grievous famine fell upon the people and pitilessly destroyed them . . . But on the very day on which the nation received the Baptism of the faith, there fell a soft but plentiful rain, the earth revived, the fields grew green again, and the season was pleasant and fruitful.

The ninth century appears to have produced only one notable drought. The next series did not begin until AD 987, sixteen years after the legendary deluge brought down by St Swithun (which is not supported by fact). In 987, 988, 992, 993, and 994 the summers were so hot that the corn and fruit dried up, and 1022 was apparently even hotter. The heat was intense and killed men and animals alike. After that there were few serious droughts before the Norman Conquest, but from 1102 to 1149 the appalling story began all over again. As early as 14 April 1114 the Thames was almost completely dry and the summer so hot that cornfields and forests were burned down. Even in the following October children waded across the river without difficulty.

This was not the end of the series of droughts. They occurred at regular intervals into the thirteenth century. In 1252, a particularly disastrous year for unrelieved heat and blinding sun, the grass was so burnt up that 'if a man took up some of it in his hands it straight fell to powder'. Diseases such as 'sweats, agues and other' were rampant.

But the Middle Ages is known as the period of North Sea storms, and though it was undoubtedly one of the wettest periods on record, as well as the wettest in the past 2,000 years, it is not always realised that the storm surges were separated by intervals of several years. Groups of wet years were known even in the eleventh century, after which they became more frequent until the middle of the fourteenth century.

During the latter part of this century the climate became more balanced and people were beginning to form definite opinions on what was proper for each season. Chaucer, in his famous Prologue to *The Canterbury Tales*, takes the opportunity to praise the virtues of 'April, with her sweet showers' and the freshness of the countryside in springtime. Had he lived a century earlier he might have had reservations about this season and been forced to see his fellow pilgrims wallowing in deep floods during one year and wilting under an arid sun in the next.

From about this time it is possible to trace an interesting change in people's attitudes towards the weather: it was no longer regarded as an everyday normality, beyond the power of anyone to alter for better or worse, but rather as either a gift or a punishment from the Almighty. Various explanations have been given to account for this, not the least being man's eagerness to use the weather as a symbol in the justification of wars. When it favoured us—for example, at the Battle of Crécy in August 1346, and when Philip of Spain sent out his two armadas—it was proof of a righteous cause; when it failed to do so it showed that, though the cause was still righteous, God was intensely displeased with the lack of true religion and the other shortcomings of the British people at home.

Alternate small groups of good and bad years occurred during the reigns of the Tudors and early Stuarts, our first Tudor king seizing the crown in 1485, not in 1483 as he had originally hoped, as in that year his armies were thwarted by the Wye and Severn floods after a ten-day struggle.

But by 1650 the climate was becoming more extreme again. From then until 1850 winters were often very severe, so much so that the period is now known as 'the little Ice Age', and no reference to it is complete without details of one or other of the famous frost fairs held on the Thames and other rivers. In his *Diary* for 24 January 1684, John Evelyn remarks:

> Coaches plied from Westminster to the Temple, and from several other staires, to and fro, as in the streetes, sleds, sliding with skeetes, a bull-baiting, horse and coach races, puppet-plays . . . so that it seemed to be a bacchanalian triumph, or carnival on the water.

The winter seasons of 1794–1810 were particularly severe. Evidence for this abounds, not only in statistics but in literature. Samuel Butler, writing in retrospect of the early years of the nineteenth century, declared:

> In those days the snow lay longer and drifted deeper in the lanes than it does now, and the milk was sometimes brought in frozen in winter . . . I suppose there are rectories up and down the country now where milk comes in frozen sometimes, in winter, and the children go down to wonder at it, but I never see any frozen milk in London, so I suppose the winters are warmer than they used to be.

After 1810 it became slightly milder for a time, but cold winters soon returned, beginning with the season 1813–14, when the last frost fair was held on the frozen Thames, and continued to be fairly regular until 1845. It is interesting to note that this period coincided with the most impressionable part of Charles Dickens' life and coloured his writings. In particular, five of the six Christmases from 1817 to 1822 were bleak in southern England, and several suffered heavy falls of snow. As these winters would have been about the earliest within the novelist's recollection, they no doubt left him with the belief that such conditions were proper to the season. The notion gained strength in the 1840s from the introduction of German greeting cards portraying snow scenes, and only in fairly recent years has the proportion of Christmas card snow scenes diminished.

After the middle of the last century cold winters became gradually less frequent, and a milder cycle began to be established. The process was so slow at first that it was hardly noticed. For the period 1891–1930 the Greenwich records show a slight increase of temperature over the 1841–90 period amounting to 1·4° F in December, 1·2° F in January, and 0·3° F in February. But the whole of the region between central Europe and the Arctic was affected, as were many parts of the New World. In Philadelphia the mean annual temperature rose from approximately 52° F in the 1830s to more than 56° F in the 1930s. At Spitzbergen, in the European Arctic, the rise between 1912 and the middle 1950s was approximately 10° F.

On the Russian coast the rise of temperature was even greater. All over the northern hemisphere glaciers retreated and the salinity of the ocean increased rapidly. Fish, including haddock, halibut, herring, and especially cod, migrated far to the north of their normal grounds, and became common in Greenland waters. In Greenland and Norway, ground previously frozen since Viking days was used once again for cultivation. The effects were heightened because they were cumulative, because the warming-up process applied primarily to the winter season, and, in addition, because the rise of temperature became particularly well marked between 1930 and 1960, with the rather odd exception of the early war years and 1946–7. And whereas, originally, the temperature rise so far as Europe was concerned began in the centre of the continent and then spread gradually northwards, from about 1930 onwards it was the Arctic region that was principally affected.

It was therefore hardly surprising that hopes were expressed that the British climate would become progressively warmer, and eventually in summer as well as in winter. But, now, the rise in the temperature of the northern hemisphere winter has been halted, and, in some places, even reversed, and the immediate prediction is that during the next few decades easterly and northerly winds will become somewhat more frequent, at the expense of the more normal south-westerlies. One result of this should be the occurrence of a greater number of snowy days over the British Isles, and the benefit that this can bring to our developing ski-ing resorts requires little emphasis. On the debit side, however, severe winters like 1946–7 and 1962–3 must be expected to occur, on average, rather more often than about once every 15 years or so. The current climatic trends should be closely followed by all interested authorities, so that provision for what is to come may be made in good time.

It is too early at present to predict a return to the winter patterns of Victorian and pre-Victorian times with any real certainty. But speculate one must, and if the Thames flowed less freely than it does today, and continued to be spanned by small wooden bridges that impeded the currents, frost fairs would still be definite possibilities during excessively cold winters. Ice flows have formed in British rivers on more than one occasion during the present century, as well as in the previous 50 years when winters were occasionally severe.

Taking the very long term outlook, we are on different ground, and we can talk only in terms of vague possibilities by examining what has happened in the remote past and assuming that the events in question could, in certain circumstances, occur again in the future. The table on p 98 shows the broad climatic eras from recent to prehistoric times, and these have varied from warm—with temperatures that are found today only in tropical and equatorial zones—to glacial. Of the three glacial periods, separated by intervals of some 200 to 300 million years, the last one is comparatively recent, dating from a mere 1,000,000 years ago. The present ice sheets of Greenland and the Antarctic are relics of this Ice Age, and it was only in about 6500 BC that the last great ice sheet disappeared from Scandinavia. Over much of Europe there then followed a change to a warm humid climate, with a heavy rainfall that caused considerable growth of peat. This period lasted approximately 4000–2000 BC and it is thought that the many stories of the Great Flood that have been handed down over the centuries date back to it. Afterwards there was a gradual decrease of rainfall with long periods of drought interrupted by temporary returns to more rainy conditions,

Era	*Formation*	*Climate*
Quaternary	Recent	—
	Pleistocene	Glaciation in temperate latitudes.
Tertiary	Pliocene	Cool.
	Miocene	Moderate.
	Oligocene	Moderate to warm.
	Eocene	Moderate, becoming warm.
Mesozoic	Cretaceous	Moderate.
	Jurassic	Warm and equable.
	Trias	Warm and equable.
Palæozoic	Permian	Glacial at first, becoming moderate.
	Carboniferous	Warm at first, becoming glacial.
	Devonian	Moderate, becoming warm.
	Silurian	Warm.
	Ordovician	Moderate to warm.
	Cambrian	Cold, becoming warm.
Proterozoic	Keweenawan	Glacial.
	Animikian	—
	Huronian	—

leading to the cold wet period that began around 500 BC, referred to at the beginning of this chapter. It is odd that such an apparently static thing as climate behaves like a spinning top when the time element is compressed.

World climates are to a large extent inter-related. Britain's present climate is often described as 'cool, temperate' because, whilst it holds sway, the prevailing wind from the Atlantic acts as a temperature regulator and prevents large-scale heating actions or the development of intense cold. In paying the price for this, we find that the oceanic windflow gives us more unsettled days than fine, for barometric pressures in the middle and northern regions of the North Atlantic are frequently low and associated with the passage of depressions. To the south, over the sub-tropics, barometric pressures are normally higher and the weather much warmer and drier. Well to the north, in Arctic regions, barometric pressures are again comparatively high and the weather dry and intensely cold for much of the year.

The seasons in Britain and in other parts of western Europe are caused not merely by the annual rise and fall of temperature associated with the earth's position relative to the sun but also to a large extent by the temporary movements of the high-pressure belts to the north and south, each of which has a marked influence on our atmospheric circulation, either directly or indirectly. However, if the central positions—and therefore the boundary regions—of these two belts depart from their normal latitudes for comparatively long periods, then we shall experience a change of climate. It will become basically colder if the northern high-pressure systems move to the south, and warmer if the southern 'highs' drift gradually to the north.

So far no one has discovered the basic causes of long-term changes of climate, but scientists consider that they are brought about either partly or wholly by variations in the amount of energy emitted by the sun and in the amount that reaches the earth. Dust or gaseous particles in the atmosphere can absorb solar energy, so that there is real ground for believing that world climates could suffer interference from such things as meteoric, volcanic, and nuclear dust, or even from carbon dioxide created by persistent large-scale coal and oil combustion and other man-made devices.

In the search for enlightenment and to prove the contention that man may have unwittingly interfered with the elements, it is easy to forget that climatic changes of great magnitude have been taking place since the earth was first formed. So it is unwise to reach conclusions too hastily. In their time, the invention of the steam engine and of radio waves were blamed as being responsible for creating floods, storms, and other unwanted excesses. Yet there is little doubt that a single change of climate, when it does occur—and no matter how it is caused—can trigger off a whole series of changes with perhaps world-wide repercussions.

In this connection it is reliably estimated that only a relatively small change in the amount of heat received at the earth's surface is required to ensure either the growth or the dissipation of polar ice sheets, and neither process is desirable if it is to be prolonged.

An annual increase of temperature at the rate experienced at the beginning of the present century would result in large-scale melting of northern ice-sheets, and this would lead to a rise in sea-levels that could inundate London, Paris, and New York by the end of the century.

However, the warming up of the Arctic having been halted, at least temporarily, one is now led to speculate in the opposite direction. A lowering of the average temperature in northern climates would lead to a similar process taking place over Britain, and if this amounted to only a matter of 2–3° F, it could be enough in the long run to set the stage for a new ice age. But the process would take hundreds—or, more probably, many thousands—of years.

7

LOCAL WEATHER

ALTHOUGH Britain is a small country it produces many differences in climate between and within its various regions, and how these vary from place to place and from season to season is well worth careful study. The choice of where we wish to live may depend upon it—especially if we are just about to retire, are not in good health, or suffer from a chronic ailment.

To a certain extent geographical location is important. For example, the South-west region has an entirely different type of climate from the South-east or from the Midlands. But what is particularly noticeable is that in some parts of the country the local climates vary from place to place, at times every few miles, so one cannot rely on any gradual transition when travelling on a direct course from one region to another.

The point is that the altitude of each district, the amount of shelter from—or exposure to—winds of varying directions, the type of soil, and the nearness of forests, rivers, or lakes, will all affect its local climate—so much so that local factors may sometimes be more significant than the particular latitude north or south. For example, Cambridge in winter is far colder than Lerwick in the Shetlands, and the average winter night temperature in the Orkneys and Shetlands is almost the same as it is along the south coast of England.

However, if, when choosing a place in which to live, one has the freedom of choice of one general region of the country or another, then broad regional climates should be considered before reference is made to any very localised records. It is often said that the British climate is very equable, without any

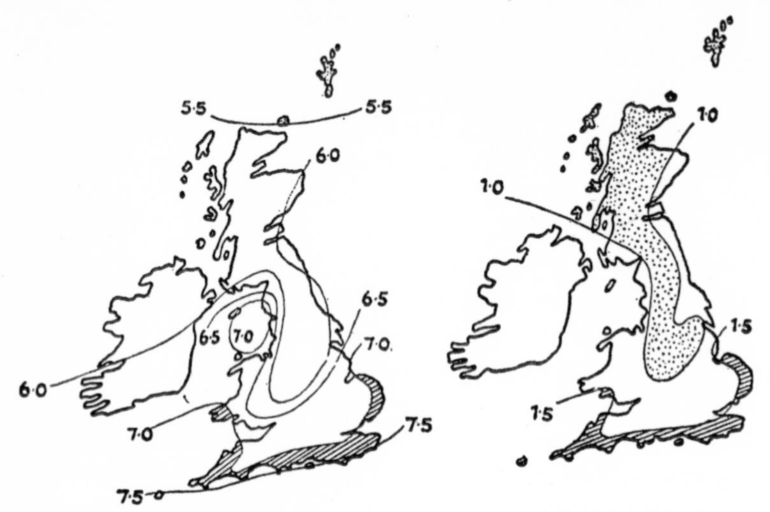

Fig. 14. The average number of sun hours over Britain: left—for the month of June; right—for the month of December.

very great extremes of temperature, but it is far more equable in the west than in the east. Southern districts are much sunnier than northern ones, particularly during the winter months. It is normally much drier in the south and east than in the west and north, and more bracing in the east and north than in the south and west.

Not only are some people more sensitive to weather than others, but many have specific requirements for (or allergies to) one or other of the various elements which go to make it up; and one must include not only sunshine, temperature, wind, rainfall, and snow, but also such factors as the annual frequency of thundery and sultry days, the varying humidity from month to month, and the frequency of fogs and frosts.

So far as sunshine is concerned, Britain fares rather better than many overseas critics try to make out. To give some figures for Europe north of the Alps, upwards of 1,900 hours of sunshine per year are recorded in parts of southern Germany, on the Baltic Islands off Sweden, and also in the vicinity of the Channel Islands. West Sussex coastal areas have just over

1,800 hours, compared to 2,000 hours on the west coast of France. South-western districts of England, with between 1,550 and 1,800 hours, are less sunny than the South-east area but sunnier than East Anglia (1,450–1,500 hours). By contrast, the cloudiest regions of northern England and Scotland have an average of only just over 1,000 hours of sun per year.

Except over the highest mountain regions in Britain, there are more hours of bright sunshine each year than of rainfall. Our sunniest mainland resorts are situated along the south-east coast, and it is interesting to note that differences in sunshine totals experienced on either side of the English Channel are negligible. But one must allow for the fact that there is about 5–10 per cent less sunshine available at inland situations, discounting the extra foggy spots, than at nearby coastal resorts. In the county of Devon, Torquay, for example, has an average of 150 hours more sunshine than the region around Exeter. Where the countryside is excessively hilly, the difference in sunshine between coastal and inland regions becomes greater; for hills and mountains produce large areas of shadow, and they also produce more low cloud than the plains.

Over the year as a whole average temperatures increase from north-east to south-west by 6° F, the figure for north-east Scotland being 46° F and for south-west Cornwall 52° F; and in most individual months the difference between the coldest and warmest regions is, on average, approximately the same.

The south-west of England has been a particularly fruitful hunting ground for new housing developments during the last 15 years, partly because a number of light industries from London and the Midlands have moved to this region, but more particularly because it is a favourite place for retirement. Yet it has to be admitted that most of Devon and Somerset, as well as the coastal regions of north-west England and south-west Scotland, have an average of twenty days of gales per year. From the Hebrides to North Wales, thence to south Cornwall (excluding the Bristol Channel) there are more than thirty days of gales per year. The strongest gusts are invariably recorded in coastal regions, which is a penalty that has to be paid by those who wish to live near the sea. By contrast, over the broad general region from Gloucestershire to Middlesex and northwards to the Welsh border counties in one direction and to the east and north Midlands in another, there is an average of only two days of gales per year.

Comparative rainfall data for various regions of the British Isles tend to be misleading. When making enquiries one is normally quoted the average amount of annual rainfall. Duration of rainfall, which varies as much as quantity, is just as important, since this affects the general cloudiness and sunshine results.

Then there is snowfall to consider. Discounting occasional freak snowstorms that occur out of season, for most regions of Britain the heaviest snowfalls are normally confined to the periods between December and March and, to the south of the Midlands, frequently to January and February only. For low-lying districts in southern England there is an average of only about two days of snow or sleet in December, but nearer twice this number in each of the other winter months. Travelling northwards, or to greater altitudes, the frequency normally increases. Balmoral, in east Scotland, expects eight days of snow or sleet in December, nine in January, eight in February, and ten in March.

Lightning fatalities are relatively few in Britain, the chance of any individual being struck in a single year being about 1 in 5,000,000. For all that, a number of people become extremely nervous during thundery weather. In a published report, covering a recent 10-year period, a survey of thunderstorms made by the Electrical Research Association shows that the average duration of each storm is about 1 hour and that they are most frequent between May and August, with the peak period between 3 pm and 6 pm. But the incidence varies greatly between the regions. As a general rule, our hottest areas have the greatest number of thunderstorms. Most of Lincolnshire, experiences over twenty days of thunder per year, while

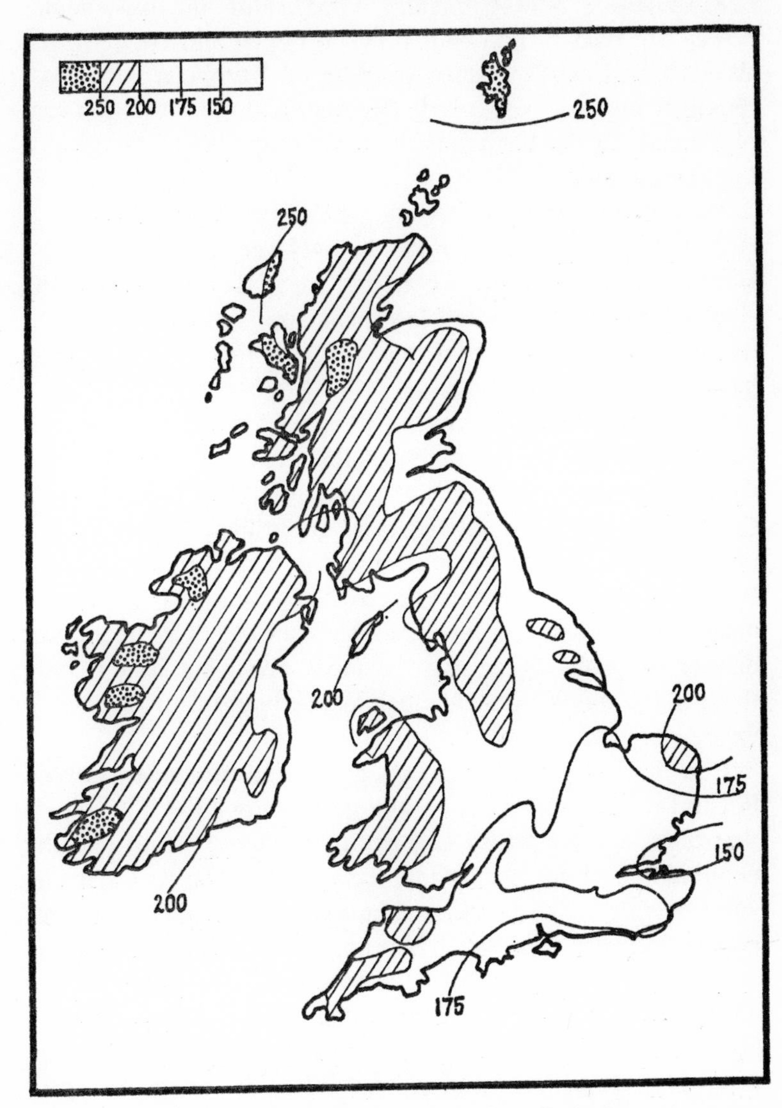

Fig. 15. The average annual number of rain days.

Leicestershire, West Yorkshire, Essex, and the mid-Thames Valley are next in line with between fifteen and twenty days. Western and north-western districts of Britain are the least thundery regions, particularly Scotland and the south and east of Ireland. In the Shetlands there are only about two days of thunder per year.

Fog, which in this country is a greater menace than lightning, is most frequent in the Midlands and over east-coast regions from Yorkshire to Kent, where fifty foggy days per year is the general expectancy. But all large industrial cities lose between one-quarter and one-fifth of their sunshine through fog or haze particles in the atmosphere.

The power of any individual to select a region in which to live is affected by other things than climate. Nevertheless, having decided on a particular county or town, it is advisable to examine the potential for local variations of climate that may occur within it. For example, the districts least affected by atmospheric pollution lie to the windward of towns, normally between south-west and north-west. Or, again, the district of one's choice may be fairly level, but if it is hilly there is the question of whether to live high up or down in a valley.

It is the wind which makes the greatest difference between hill and valley climates, for a hilltop is exposed to all the winds that blow, while a valley is generally sheltered. Wind aids the ventilation and, in moderation, has a tonic effect, but a strong wind or gale in an exposed situation is something that most people find very depressing. Not only does it have an irritating effect upon the nerves but it increases the penetrating power of rainfall and cold air. On the other hand, a valley situation also brings its problems. It is hardly likely to enjoy unobstructed sunshine, and it may also be prone to frost, particularly if it is in any way enclosed. On still nights cold air always gravitates downwards. Closed-in valleys will collect the cold air, and some may have temperatures as much as 30° F lower than on the surrounding hilltops.

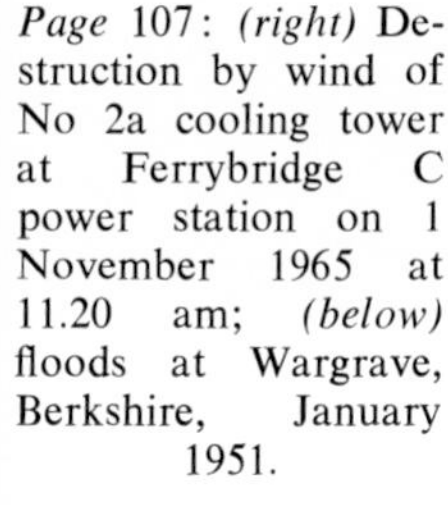

Page 107: *(right)* Destruction by wind of No 2a cooling tower at Ferrybridge C power station on 1 November 1965 at 11.20 am; *(below)* floods at Wargrave, Berkshire, January 1951.

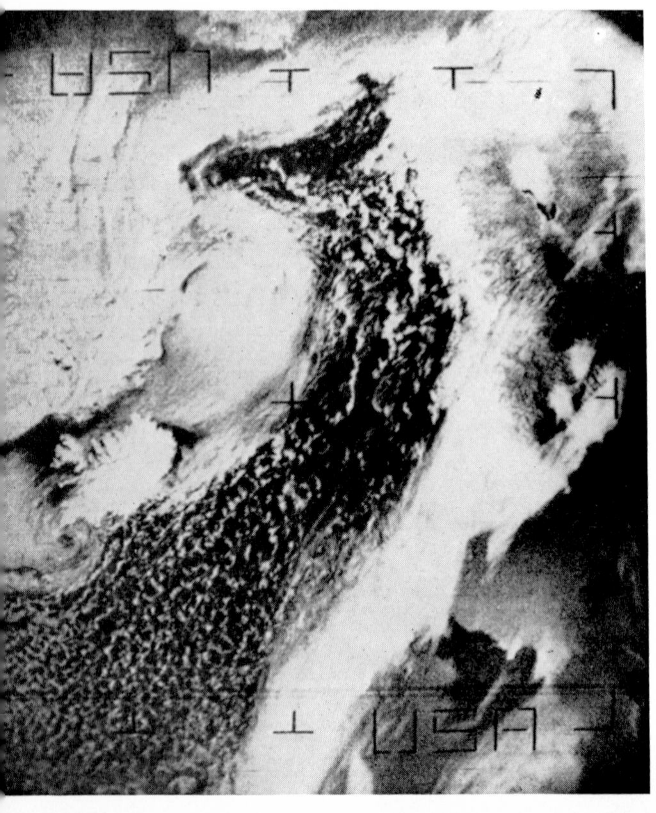

Page 108: *(above)* View of Meteorological Office headquarters building, Bracknell, Berkshire; *(left)* picture taken by ESSA 6 satellite at 11.30 am GMT, 28 March 1968, showing British Isles in lower right-hand corner. Glitter on the North Sea points to light winds. The cloud system of a depression's stationary front extends south-westwards from Norway.

Snow tends to lie in thicker strips and to last longer on hilltops than on low ground; and hill roads, especially when sunk below the general level of the ground, are readily blocked by snow. During any severe winter spell, even if the season as a whole is not unduly cold, hill villages may be isolated for some time by deep drifts of snow piled up by the wind. This was demonstrated by the cold spell during the second week of January 1968.

The most favourable situation, where the country is undulating, is neither on the surrounding hilltops nor near the valley floor, but somewhere between—high enough to be above the frost-line but sufficiently sheltered to avoid the worst winds. Even so, there are other factors to be considered, such as the advantages and disadvantages of slopes facing north, east, south, and west. The important points are the slope of the ground and the elevation of the surrounding hills in different directions relative to the site under consideration. Any slope in latitude 55° N from which the ground slopes upward to the south by a gradient of 1 in 8 will have no sunshine on the shortest day, and even if the slope is less steep than this it will probably not be a good place to live during the winter.

When choosing a home it is a good plan to consider at which time of day one most requires sunshine. Presumably the house or bungalow will have a garden, and as spring and summer evenings are times to be out in the garden, exposure between west-south-west and north-west is an advantage. A steep downward-facing south slope receives more heat from the sun than does a level site and much more than one that faces north. This may not be an advantage, however, for although it will produce early flowers and vegetables, it is likely to become hard-baked during dry weather, particularly if the soil has a marked clay content. Professor R. Schulze of Hamburg has shown that a slope of 45 degrees to the south receives 60 per cent more heat from the sun than a horizontal surface.

A number of people prefer an easterly aspect, because

although this means that their site is open to the coldest winds of the winter, it warms up quickly on summer mornings. But some shelter is desirable to protect one's property from the full force of the cold easterly winds of winter and early spring. A distant hill-range or a less distant tall building or line of evergreens may be all that is needed.

Soil differences affect local climate mainly through their different reactions to the heat of the sun and their different heat capacities and conductivities. In clear, calm weather the air above light sandy soils tends to have considerably lower night temperatures than the air above heavy clayey soils, and for this reason sandy soils are more subject to frost. On these the growing season for flowers and crops may be shortened by between six and eight weeks. But, by day, sandy soils tend to have temperatures a few degrees higher than those that are mainly of clay. However, when the soil is saturated with water, the day-to-night temperature variations become smaller. Undrained soils over clay, or clay itself, can only dry by evaporation, and these tend to remain cold and damp for longer periods than the soils through which water can penetrate and drain away underground.

The effect of the nature of the soil upon local temperatures is naturally modified to some degree by the vegetation that covers it. Grass, especially long grass, traps a great deal of air among its stems and roots, and because of this the top level of the grass is hotter by day and colder by night than is bare soil or more open vegetation. This also accounts for the fact that on cold nights shallow mist or fog layers usually develop first over grassland.

Forests have a different effect. By day the upper surfaces of the leaves take up the sun's heat and the warmed air at this level is swept away by the breeze so that the air beneath remains cool. At night the crowns of the trees cool the air that surrounds them and the chilled air then sinks. But the fall of temperature at the forest floor is less than over grassland.

Wherever there are hills and valleys, local day and night breezes will be set up on otherwise calm days. On a sunny morning the floor of an open valley warms up before the surrounding hills, which results in an up-valley flow of wind in the form of a breeze. This has an invigorating effect, which is particularly noticeable to dwellers who live some distance above the lowest part of the valley, but a cool downslope breeze will develop during the night (see p 259). Therefore, when siting a house on such a slope, it is desirable to avoid having large windows on the ground floor facing uphill. Of course, the cool night breeze, so dreaded by gardeners during the critical late spring months, can be a positive advantage in summer after a long hot day, providing relief from the heat and promoting sleep.

The conditions depend to a large extent on the direction of the valley. In a valley that opens to the south the front rooms of a house facing this direction have the advantage of sunshine and a breeze by day while the less important rear of the house bears the brunt of the night winds. A house that faces down a valley opening to the east gets the morning sun and earlier relief from the cold night winds, due to the fact that temperatures in the early morning will rise fairly quickly. But the sun goes behind the hills earlier in the evening and, as a result, night breezes set in sooner. In a valley opening to the west the reverse situation applies: evening temperatures are higher but mornings are colder. The narrower a valley happens to be, the greater will be the effect of night and day breezes.

In a forest clearing wind strength is naturally less than in the open, the sheltering effect of the trees extending downwind for at least ten times their height. The air is fairly turbulent in the clearing, so that smoke from chimneys dissipates quickly. Another possible advantage of such a position is that temperature variations are smaller than in the open, but this varies considerably according to the type of tree which makes up the forest. Light conifers such as pine and larch offer less resistance to the sun in summer than do well-grown deciduous trees—but more in winter. Consequently the daily range of temperature in a plantation of light conifers is greater than in a deciduous forest in summer but less in spring and autumn. On the other hand, a dense thicket of spruce forms an almost impenetrable obstacle, shutting out the sun, wind, snow and, to some degree, even rain.

For those who suffer from hay fever, trees in the neighbourhood are a disadvantage as a result of pollen in the air, especially in the early part of the year between mid-January and June. The main sources of pollen among trees are ash, oak, and elm that are within a quarter of a mile or so, but a much greater amount of pollen is provided by grasses, especially in the early summer. The most favourable situation for those who suffer from hay fever is near the coast where prevailing winds are onshore.

The effect of a lake near a house depends on the depth and movement of the water. A large sheet of deep water has a moderating influence on temperature, for on a hot summer day it cools the air blowing over it and on calm days helps to give rise to a gentle breeze that blows from the water to the shore. On cool nights the opposite effect is found. The water, which is then warmer than the land, heats the air that blows over it. This air rises and draws in air from the land, thus setting up a gentle breeze from shore to lake.

A sheet of deep water has another advantage in that it helps to reduce the risk of frost. However, shallow water is less effective in this respect and it also encourages flies.

Sizeable rivers act like deep lakes in moderating the extremes of temperature, but the choice of a place in which to live, particularly if it is up-river, must take into account the highest flood levels that have occurred in the past. More than just a few people have found to their cost that it is one thing to have a river at the bottom of the garden, but quite another to have the garden at the bottom of the river.

Living near the sea presents advantages and disadvantages

in almost equal proportion. The fresh breezes and clean air are helpful factors, and sunshine near coasts is normally higher than in inland areas. However, the wind will become excessive at times, particularly during the autumn and winter months on southern and western coasts. On the low-lying parts of the east coast the risk of storm floods is one that must always be taken into account, and some stretches of coastline are rather prone to sea fog or mist.

Most local districts, where the ground is undulating, have places that are more vulnerable to lightning than others. The highest points are normally the most dangerous, but one must also bear in mind that electric storms often persist much longer over river valleys than elsewhere. Sometimes they run up and down a valley for some hours before dispersing. Individual large trees in exposed positions are often the first to be struck, and oaks are five times as vulnerable to lightning as any other tree.

More often than not the weather patterns over Britain are changing frequently, so that some days feel particularly bracing, others relaxing. But some places are more bracing or relaxing than others—a factor to be taken into account when choosing where to live. For example, for those who prefer a bracing climate, towns such as Bournemouth or Torquay will not be suitable. The people in question will constantly complain of feeling tired and wanting to go to sleep.

As a rule, bracing or 'tonic' local climates occur where there is plenty of sunshine, dry breezy air and moderately low average temperatures, and a fairly large temperature range between day and night. These climates are particularly good for people with a tendency to tuberculosis and for recovery after moderate overwork. They are also good for healthy growing children, as they encourage them to spend much time in the open air. But they are less suitable for those who suffer from rheumatism, skin troubles, or nervous diseases. Relaxing or 'sedative' climates, which have greater cloudiness, higher humidity, and more equable temperatures, are suitable—provided they are sufficiently sheltered—for those who have respiratory, heart, or kidney troubles, and for old people, very young children, and delicate subjects of all ages.

Highly bracing climates in Britain are mainly limited to the north-east coast and to the High Peak of Derbyshire. Bracing climates include the rest of the east coast, a good deal of the Sussex coast, the coast of north-east Cornwall, and hilly districts generally. The main areas of relaxing climate include the south coasts of Devon, Cornwall, and Wales, and most of the low ground in the western half of England.

But, in addition, much depends on the aspect of individual locations. Sidmouth in South Devon, for example, is less relaxing than many nearby places, because it is open to breezes from the south-east. For this reason those who contemplate moving to a new district and are anxious about the climate should make full enquiries from the appropriate local authority regarding what is commonly called 'the property of the air'.

8

WEATHER LORE

DURING recent years there has been a growing interest in weather lore and in the old sayings and rhymes on the subject that have been handed down from one generation to another over many centuries. Not all the weather sayings one hears today in this country actually originated in Britain, and that very famous one about red skies at night and in the morning seems to have been imported at some time. For in *Matthew* xvi, 2, 3 Christ declares:

> When it is evening, ye say, It will be fair weather: for the sky is red. And in the morning, It will be foul weather today: for the sky is red and lowring.

References to weather are frequent in the Bible. 'God prepared a vehement east wind', we read in *Jonah* iv, 8. Similarly, in *Ezekiel* xxvii, 26, 'The east wind hath broken thee in the midst of the seas'. In *Acts* xxvii, 14 there is reference to a tempestuous wind called Euroclydon, which is a strong east-north-easter. 'And cold out of the north', states Job (*Job* xxxvii, 9). 'A whirlwind came out of the north', is how Ezekiel put it (*Ezekiel* i, 4). Cold northerly winds have the ability to penetrate, with accompanying snow showers, squalls, and occasional whirlwinds, to many parts of the northern hemisphere, and this provides just one example of why northern Europeans may feel familiar with weather conditions existing in different lands which, perhaps, they have never visited. It also shows how nations can share a common heritage of weather lore despite climatic differences that exist between one part of a continent and another.

On this subject, it is interesting to note that, in 1949, a scientific 'jury' in the United States found that a number of European weather proverbs were valid for North America, too, and 87 of 153 sayings examined were found to be true in terms of scientific principles. Included in the group of reliable sayings was

> Red sky at night, shepherds' delight;
> Red sky in the morning, shepherds' warning—

but it requires elaboration. An evening sky which is more of a rosy shade than a dark red and which occurs without too much cloud thickening towards the horizon is usually a fair weather sign. The colour is caused by the sun's light being bent by haze particles in the atmosphere, and it is during the evenings that haze accumulations are usually greatest. But an angry deep red glow under a heavy layer of cloud, frequently seen at the beginning of a stormy day, will signify rainy weather even if it occurs at night. The colour in this case is caused by sunlight passing through water droplets in the atmosphere.

Another saying reveals that

> The evening red and the morning grey
> Are the tokens of a bonny day.

The morning grey is early morning mist. During late autumn and winter the mist may be slow to clear, but at other times it is normal for the sun to clear the mist by about 10 am, and the remainder of the day is usually fine.

Other sky and cloud colours that have value in forecasting the weather include purple, which is normally associated with haze and indicates mainly settled weather; green, which is rare but may now and again be momentarily seen under a heavy cloud layer just before the onset of rain; and yellow, which is a sign of an approaching rainbelt and increasing wind from a point between south-west and south-east.

One will nearly always find some proverbs quoted in the old weather almanacs that are based purely on superstition, one example being the following piece of Saxon lore which held that a year's weather was influenced by the day of the week on which the first of January fell:

> *Monday:* a severe and confused winter, a good spring, a windy summer; *Tuesday:* a dreary and severe winter, a windy spring, a rainy summer; *Wednesday:* a hard winter, a bad spring, a good summer; *Thursday:* a good winter, a windy spring, a good summer; *Friday:* a variable winter, a good spring and summer; *Saturday:* a snowy winter, a blowing spring, a rainy summer; *Sunday:* a good winter, a windy spring, a dry summer.

Others, seemingly not so far-fetched, still attempt to link one season with another. There is the warning

> If the ash is out before the oak,
> You may expect a thorough soak.

But if the subsoil is moist the oak is always in leaf before the ash. So much for this clever little rhyme, which comes from Kent. Another rhyme, which originates in Shropshire, says that 'When the ash is out before the oak, then we may expect a choke', and 'choke' is the old word for drought.

Another ancient belief is that when the autumn hedgerows are decked with an abundance of hips and haws, or when holly berries are plentiful, this indicates a severe winter to follow, for kindly Nature provides the birds with an extra large store of berries if, later on, the ground is to be long frozen or covered with snow, thus creating a scarcity of insects and other food. But, in fact, it is the weather of the preceding spring that determines whether berries shall be plentiful or not.

Then there is the saying

> March comes in like a lion
> And goes out like a lamb.

This happened in 1965 in many parts of Britain, but in some years the reverse occurs. Or the month can begin with storms and end with storms. The trouble with so many of these sayings is that they were produced in the first place by inaccurate observation or, as Dr Johnson put it, by the human tendency to represent 'as constant what is really casual'.

But there are more than just a handful of weather proverbs that express a basic truth or a near truth in a fanciful, often figurative manner. The Swithun legend, for example, alleges that the Saint brought down a forty-day deluge in protest at having his remains transferred from an open grave to the inside of Winchester Cathedral on 15 July 971. There is no historical evidence for a flood or a long wet spell at this time but, somehow, the question still arises: can Britain reckon on receiving forty days of rain if even so much as a shower occurs on 15 July? Records show that this is far from being the case, but it is true that a long wet spell during the early part of the summer tends to be followed by further periods of wet weather during late July and August.

Not only in Britain but in many parts of Europe, saints' days in particular seem to have been selected as exerting a special influence on the weather, and the St Swithun legend has counterparts in a number of European countries. In each case the saint in question, the tradition or myth attached to him, and the dates of his supposedly taking charge of the weather, are subject to variation, yet the period of the alleged influence is a fairly consistent forty days. This suggests that all these tales had a common origin in the world-wide tradition of Noah's flood.

In the north of Scotland St Martin Bullion (Martin 'le Bouillant') has his anniversary celebrated on 4 July. In France there is the feast-day of St Benedict (St Benoît) on 21 March, St Médard on 8 June, St Protase on 19 June, and St Anne on 26 July. All these saints have Swithun-type attributes, and it is said that, many years ago, soldiers of the French army, sorely harassed by heavy rainfall during a summer campaign, broke into several churches dedicated to St Médard and took their vengeance on him by smashing the altars.

To complete the list, Belgium's rainy saint is Godiève, whose date is traditionally 27 July but, according to the official calendar of the Roman Catholic Church, 6 July. Italy has St Bartholomew (24 August) while, for Germany, the rainy influence is attributed to the Seven Sleepers of Ephesus, who share their date with St Godiève, across the Belgian frontier, on 27 July.

'Where the wind is on Martinmas Eve (10 November),' runs an oft-quoted piece of prophecy, 'there it will be for the coming winter.' This is just another example of the many sayings that encourage the credulous to believe that a single day's weather is responsible for several months of ensuing conditions. 'As November 21st, so is the winter'; 'As at Catherine (25 November) foul or fair, so will be the next February', quotes the Richard Inwards' collection of weather sayings.[1] Another predicts 'for every fog in October a snow in the winter, heavy or light according as the fog is heavy or light'.

As with the story of Swithun's forty days of rain, there are times when sayings in this class come fairly near to hitting the truth, but, not surprisingly, this is rather exceptional, and it would seem that few of the devotees of this type of weather saying have carried out statistical analyses to prove their favourite contentions.

Attempts have however been made to verify the following assertion:

> So many fogs in March you see,
> So many frosts in May will be.

Mr E. L. Hawke, a former vice-president of the Royal Meteorological Society, declares that this is a saying often fulfilled even if it does not mean that the two events are directly linked. During March and May respectively the seasons for foggy weather and frosty weather are drawing to a close in southern England, the one being just about as infrequent in the former month as the other is in the latter. Hawke found that in London there are four foggy mornings in an average March and four nights of ground frost in an average May.

But other researchers have compared the March and May records for some 200 British weather stations, and for every one that showed an equal number of fogs and frosts there were two that showed more frosts than fogs. The areas that approached equality in the frost-fog ratio were mainly industrial, where the frequency of fog is considerably augmented by atmospheric pollution. In most of the non-industrial areas there were at least two May frosts for every March fog.

In the saying

> If there's ice in November that will bear a duck,
> There'll be nothing after but sludge and muck

there appears to be a good deal of sound commonsense, and one can take November 1965 as a case in point. A bitterly cold spell, with east winds, heavy snowfall, and zero or sub-zero temperatures, affected the whole of the North Sea region, including Scotland and northern England, between 12 and 16 November, and, for good measure, gales brought end-of-month snowfall as well. But the 1966 winter was mild—with plenty of 'sludge and muck'.

A number of reliable weather sayings are to be found in the group that interpret the behaviour of birds and insects, yet only a few of these can be proved scientifically. Very small creatures appear to be sensitive to minute changes in the atmosphere that precede what we regard as actual changes in weather; to us therefore they may justly be regarded as prophets even if, for their own part, they are only responding to existing conditions. Perhaps one day the ornithologists will account for the mysterious action:

> When rooks seem to drop in their flight,
> as if pierced by a shot, it is said to foreshow
> rain;

and why

> If the wild geese gang out to sea,
> Good weather there will surely be.

In the same category is the saying, 'Bees will not swarm before a near storm', which appears to be true enough.

Take, now, the off-quoted saying

> Swallows high,
> Staying dry:

Swallows low,
Wet 'Twill blow.

The explanation is simply that swallows are able to catch insects at high levels during periods of fine weather, but when a change is about to take place insects are found only at lower levels.

Larks are said to fly high and sing long when the weather is settled, but

If the cock goes crowing to bed,
He'll certainly rise with a watery head.

On the other hand, it is said that if a cock crows during a downpour it is a sign that a clearance in the weather is imminent.

From all too frequent personal experience most people would agree that rain is indicated within a few hours, if not sooner, when seabirds appear in flocks on the mainland and do not settle.

Cattle and horses are troubled very greatly by flies during humid weather, particularly when the weather is warm and a rainbelt is approaching Britain from the Atlantic. So there is justification for the saying: 'When cows slap their sides (or a hedge) with their tails, it is a sign of rain'.

Flies will annoy humans, too, during periods of steadily increasing summer humidity, as a rainbelt approaches. The appropriate weather rhyme puts it—quite truthfully—thus:

A fly on your nose, you slap, and it goes;
If it comes back again, it will bring a good rain.

Dogs have been known to become restless and even howl before the approach of a storm, but one must be familiar with a particular dog to be able to put the correct interpretation on its behaviour. More obviously restless at the first indication of a change in the weather are cats, which seem either to treat the matter as a joke and engage in excessive play or chair-scratching or else move jerkily and appear generally keyed up. Donkeys, too, are good at seeing future weather. As the couplet informs us:

Hark! I hear the the asses bray;
We shall have some rain today.

But not only do donkeys bray at the approach of rain; they continue to pronounce their regular warnings at intervals of a few hours until the weather clears.

Other sayings linking the behaviour of animals with changes in weather and which are considered to be generally reliable include the following:

Rooks

If rooks feed in the streets of a village, it shows that a storm is near at hand.

Rooks will not leave their nests in the morning before a storm.

Blackbirds

When the voices of blackbirds are unusually shrill, rain will follow.

Woodpeckers

When woodpeckers are much heard, rain will follow.

Fowls

If the fowls huddle together outside the henhouse instead of going to roost, there will be wet weather.

Owls

Screech-owls are most noisy just before rain.

An owl hooting quietly in a storm indicates fair weather.

Birds whistling, and silent

If birds begin to whistle in the early morning in winter, it is a sign of frost.

When birds are unusually silent in the day-time during spring and summer, it indicates thunder.

Ponies

Capering and scampering of wild ponies is a sign of rain, as it is also when they leave their moorland lairs and come down in droves to the low ground.

Cattle

So long as cattle remain near hill-tops, there is fine weather to come.

When cattle lie down in the open during light rain, it will soon pass.

Spiders

Before rain or wind spiders fix their frame-lines unusually short. If they make them very long, the weather will stay fine for a day or more.

When the spider cleans its web, fine weather is indicated.

If spiders break off and remove their webs, the weather will be wet.

If the spider works during rain, it is an indication that the weather will soon be clear.

If the spiders are totally indolent, rain generally soon follows.

Bees

Bees are restless before thundery weather.

When bees to distance wing their flight,
The days are warm and skies are bright;
But when the flight ends near their home,
Then rain and cold are sure to come.

A bee was never caught in a shower.

There is a saying about dew:

With dew before midnight,
The next day will sure be bright

and another saying tells us that 'If on a clear summer night there is no dew, expect rain next day'. Dew, as it happens, is a by-product of fine weather, not the cause of it, but since, during a fine spell, it is quite common for several dewy nights to occur consecutively, it is hardly surprising that people regard dew as a bringer of sunshine.

Visibility at all times can give a vital clue to current weather conditions. During periods of settled weather, it tends to deteriorate because most of the haze in the lower sections of the atmosphere stays within 1,000 ft of the ground level under these conditions.

Nevertheless, distant objects may appear very well defined at the beginning of a settled spell, at least for a day or two. It is therefore unwise to accept without reservation the saying, 'the farther the sight, the nearer the rain'. But if distant hills or other objects appear to be closer than they really are, then, in Britain, this is invariably a sign that rain will occur within 6 or 7 hours.

Seeing that rain, or the cessation of rain, is one of the most popular topics of conversation in this country, it is hardly surprising that many proverbs of British weather lore are concerned with rainfall more than anything else. One that sometimes works out quite well for most eastern areas of the country but which may be less applicable to the rainier west of Britain declares, plainly enough:

Rain before seven,
Clear by eleven.

It is based on the fact that there is a limit to the length of rainfall that is produced over any one place by the fronts of a travelling low-pressure system, and if it has been raining during the previous night it is customary to witness an improvement before midday. On the other hand, when one takes the narrow interpretation that rainbelts do not normally produce more than 4–5 hours of continuous precipitation—the chance of the saying proving to be accurate is less.

Since the prevailing winds over Britain are south-westerly, most of our rainfall comes from this direction as well. It is rare for easterly winds to give much rain, though they quite often bring cloudy conditions to North Sea coastal areas. Yet

When the rain is from the east
It lasts a day or two at least.

This curious situation is explained by the fact that Atlantic depressions sometimes take a more southerly course than normal and move in the direction of the Mediterranean instead of towards Scotland, Iceland, or Greenland. The result of this is that much of Britain, and, in particular, Southern England, comes under the influence of easterly winds in the area to the immediate north of the depression; in addition, the low-pressure centre is close enough to give long periods of continuous rainfall. Not until the depression weakens—it is normally slowing up at this stage of its development—will there be any marked improvement in the weather, and a day or two of steady downpour is not unusual in the circumstances.

Page 125: *(above)* Forecaster drawing up a chart at the London Weather Centre; *(below)* the London Weather Centre in Kingsway.

Page 126: *(above)* Cossor CR353 windfinding radar. Ground monitoring instrument is seen on left and balloon-carried equipment on right, balloon at point of launch; *(left)* coastguard station, with northerly gale cone hoisted.

Rain coming in shower form with a strong north-westerly wind is a sign that hail is also likely. If hail occurs towards the end of a period of continuous rainfall, then a break in the overcast is imminent, and it will also become colder. Sunny periods will develop, more particularly in inland areas and to the leeward of high ground, but there will also be showers, some again being of hail and with more than just an even chance of producing lightning as well.

Other sayings concerning rainfall and well worth putting to the test are:

Rain becoming sharper

A sharp shower of rain following a period of light—but continuous—rain or drizzle is a sign that the weather will soon improve.

Sudden, and long warning

Sudden rains never last long; but when the air grows thick by degrees, and the sun, moon, and stars shine dimmer and dimmer, then it is likely to rain for four or five hours, and sometimes even longer.

Light showers during a drought

If very light, short showers come during dry weather, they are said to 'harden the drought' and indicate no change.

There is a saying, 'three foggy nights in a week, then expect foggy days as well'. This is true, for the effects of fog are cumulative when it forms on quiet days in autumn and winter. However, when it hangs below trees, it is followed quite frequently by rain; and a very damp fog or mist, accompanied by wind, means rain to come and a certain improvement in visibility.

The appearance of a rainbow is often regarded as a sign that the weather is improving. While there is not a great deal of justification for this belief, it is true that one may be seen just after a rainbelt has cleared the area, but showers may follow. Further rainbows are then likely to be seen between the showers and are known in some districts as 'weather-galls' or 'wind-galls'. A rainbow seen in the early morning, particularly if it lies to windward, is a sign of imminent rainfall. Sometimes the prismatic colours of the sun may be seen against the side of a cloud rather than in the form of a rainbow, and this has long been known by mariners to be a sign of rain squalls with the prevailing winds blowing strongly from the west or north-west.

The sound of thunder usually indicates rainfall somewhere within the area, but not necessarily over the point of observation. However, 'after *much* thunder, *much* rain'—and in this case, one can be sure that the rainfall will be fairly general.

The following sayings about thunder are generally reliable:

Thunder from south and north

Thunder from the south or south-east indicates long storms; from the north or north-west, short storms.

Thunder morning and noon

When it thunders in the morning, it will rain before night. Thunder in the morning denotes winds; at noon, showers.

Thunder and lightning in summer

Thunder and lightning in the summer show
The point from which the freshening breeze will blow.

Lightning from one or more directions

Lightning signifies the approach of wind and rain from the quarter where it lightens; but if it lightens in different parts of the sky, there will be severe and dreadful storms.

One of the fascinating aspects of the study of weather lore is that what at first seem the most improbable pieces of guidance may prove in the end to be quite valuable. For example, drains, ditches, and dunghills are frequently more offensive before rain, and at the approach of a distant rainbelt, before there is any real wind perceptible, dust or loose sand may whirl round in small eddies. It may be some five hours or more after this before there is a marked increase in wind strength and the rain begins to set in.

In a similar class is the following:

When you observe smoke from the chimney of a house or cottage descend upon the roof and pass along the eaves, expect rain within six hours. Smoke rising vertically is, conversely, a sign of settled weather.

There is an old saying, 'It's too cold to snow'. This is true in a sense, because the colder the air happens to be, the less moisture it can contain, either as rain or snow. But all too often a cold dry spell gives way to snowfall, because humidities increase rapidly with the advance of south-westerly winds in a temporary retreat of the easterlies.

Will it be anglers' weather? If during damp rainy weather fish bite readily and swim near the surface, an improvement is likely, or, if it remains cloudy, it will be quiet rather than windy.

On occasions fish have been known to bite before rain, but this is more usual after a period of rainfall than beforehand, as rain reoxygenates the water, energising the fish which have remained dormant. One thing is certain: fish seldom bite when a major change in the weather is impending.

The following sayings also deserve careful note:

Pike

When pike lie on the bed of a stream quietly, expect rain or wind, and, in winter, cold weather.

Trout

When trout refuse the bait or fly,
There ever is a storm a-nigh.

Clam-beds

Air bubbles over the clam-beds indicate rain.

Porpoises

A school of porpoises in or near harbour is a sign of strong winds or storms approaching.

Sea-anemones

The sea-anemone closes before rain, and opens for fine, clear weather.

Frogs

When frogs croak much, it is a sign of rain,
The louder the frog, the more the rain.

Grass Snakes

An abundance of grass snakes is a sign of rain. They will also be seen nearer to houses before rain.

Worms

Worms descend to a great depth before either a long drought or a severe frost.

Over the centuries country people have been known to regard plant behaviour as an indication of future weather conditions. Some plants do make reliable guides, but very often they give only very short notice of what is to come. Chickweed will expand its leaves boldly and fully when fine weather is to follow. The scarlet pimpernel, a more popular prophet and often known as 'the poor man's weather-glass', when seen to close its flowers in the day-time, 'betokeneth rain and foul weather: contrariwise, if they be spread abroad, fair weather'. 'Goat's beard' keeps its flowers closed in damp weather, and dandelions will close quite often before a storm. If the down flies off 'colt's foot', dandelion, and thistles, when there is no wind, this is a fairly reliable sign of rain. Pondweed will generally sink before the approach of rain.

Leaves of certain trees are good thunderstorm predictors. The silver maple shows the lining of its leaves before a storm; likewise the leaves of the lime, sycamore, plane, and poplar. The trembling of aspen leaves in calm weather indicates an approaching storm. Dead branches falling in calm weather, particularly during the autumn, indicate rain.

A leech in a jar was one of the chief meteorological attractions at the Great Exhibition of 1851. A cord, tied loosely to the leech, extended from the jar and rang a little bell above whenever the creature moved. The instructions ran as follows:

> The leeches remain at the bottom during absolutely fine (and calm wet) weather. When a change in the former is approaching, they move steadily upwards many hours, even twenty-four or rather more, in advance. If a storm is rapidly approaching, the leeches become very restless, rising quickly; while previous to a thunderstorm they are invariably much disturbed, and remain out of the water. When the change occurs and is passing over, they are quiet,

and descend again. If under these circumstances they rise and continue above water, length or violence of storm is indicated. If they rise during a continuance of east wind, strong winds rather than rain are to be looked for.

Much is said about the effects of the moon on the weather. Today, as in the past, there are those who hold that lunar and weather phases go together, but statistics show that the relationship is a fairly limited one and is not strong enough to support most of the old beliefs that one hears. As one rhyme puts it:

If we'd no moon at all
And that may seem strange,
We still should have weather
That's subject to change.

In any case the changes of the moon are regular, but those of the weather—at least in Great Britain—are not. What cannot be argued is that, apart from controlling the tides, the moon acts as a very useful aid to professional and amateur forecasters alike by spotlighting any cloud formations that are present. And:

If the moon rises haloed round,
Soon you'll tread on deluged ground.

The halo—or a pale or 'watery' moon, or one with a blurred outline—indicates the presence of a *cirrostratus* or *altostratus* cloud sheet, which, in fact, is the advancing edge of a warm front rainbelt. It will be approximately 5–8 hours before rain begins to fall. Solar haloes or a watery or blurred sun foreshadow a similar development.

Very often, between rainy days but when it is still generally cloudy, the weather becomes calm, and one can hear faint sounds at an unusually great distance. This is a sign that the weather will not remain still for much longer. As the appropriate saying tells us: 'A good hearing day is a sign of wet'. Again, a murmuring—some call it a 'roar'—that is heard in a wood or forest when, outside it, there is little feeling of any wind, is a sign of rain and wind to come.

Some people, whether they know it or not—and they generally do—have a weather forecasting mechanism built into them. If they are inclined to be of a nervous temperament, they may find that they have a sense of dread or feel depressed before a fall of rain. With some persons, the feet tend to go cold before snow, and the blood vessels relax when it falls. Other people say that they can tell when the weather is changing because their ears ring, and this gives them adequate warning. Even dreams are said to be of assistance in foretelling the weather, and it is alleged in some of the old almanacs that 'dreams of a hurrying and frightful nature' are a sign of a sudden, probably a violent, change in the weather; likewise imperfect and fitful sleep. But no doubt there are too many causes behind these occurrences to make their prognostic weather value of any real use.

Frost, or the lack of it, can be useful in forecasting. If frost occurs immediately after a period of continuous rainfall, it will not last for long. But a gradual transition from rainy to more settled weather during the winter months will lead to a fall in temperature and to the regular appearance of frost if, previously, the wind has been blowing from a direction between north-west and north. If it then veers to north-east or east, the frost is likely to be fairly prolonged and possibly severe. There will also be fog at times, whenever the wind falls light or gives way to calm.

Among signs indicating the end of a frosty spell of weather are the sun looking watery at rising or setting, the appearance of high clouds driving up from the south, the stars looking dull, and the moon's outline appearing blurred.

Without doubt, the most immediately useful sayings in any anthology of weather lore are concerned with the shape, movement, colouring, and general development of cloud formations. Two of these were quoted in Chapter 1, and among others that have a good record of reliability are the following:

Small and white, like fleeces

If woolly fleeces spread the heavenly way,
Be sure no rain disturbs the summer day.
(This is a reference to small *cumulus* clouds of fair weather.)

Like a man's hand

Behold, there ariseth a little cloud out of the sea, like a man's hand . . . And it came to pass that the heaven was black with clouds and wind, and there was a great rain (1 *Kings* xviii, 44, 45). (The reference is to the top of a large shower cloud which, in the distance, appears to be small.)

According to colour and outline

Light, delicate, quiet tints or colours, with soft, undefined forms of clouds, indicate and accompany fine weather; but unusual or gaudy hues, with hard, definitely outlined clouds, foretell rain, and probably strong wind.

Capping hills

When mountains and hills appear capped by clouds that hang about and embrace them, storms are imminent.

Bank in West

A bench (or bank) of clouds in the west means rain.

Stationary, piling up

When clouds are stationary and others accumulate by them, but the first remain still; it is a sign of a storm.

Settling back

When a heavy cloud comes up in the south-west, and seems to settle back again, look for a storm.

Soon collecting

If the sky, from being clear, becomes quickly fretted or spotted all over with bunches of clouds, rain will soon fall.

Curdled

A curdly sky
Will not leave the earth long dry.

'Painter's brush', Goat's hair', 'Mare's tails'

Trace in the sky the painter's brush
Then winds around you soon will rush.
(This is a form of high *cirrus* cloud called 'goat's hair' or 'mare's tails', foreboding wind and rain within about 8 hours. It appears in tufts bunched closely together, and is not to be confused with the threadlike high clouds that denote fine weather.)

'Hen's scratchings'

Hen's scarts (scratchings) and filly tails
Make lofty ships carry low sails.
(This is the same cloud formation as the previous one.)

High sheet, gloomy

A high sheet of cloud spreading across the whole sky, and casting a general gloom over the countryside, pressages rain and wind.

Two layers

If two layers of cloud appear in hot weather to move in different directions, they indicate thunder.

If, during dry weather, two layers of cloud appear moving in opposite directions, rain will follow.

Dark and heavy

Dark, heavy clouds, carried rapidly along near the earth, are a sign of great disturbance in the atmosphere from conflicting currents. At such times the weather is never settled, and rain extremely probable.

Early disappearance

If at sunrising the clouds are driven away, this denotes fair weather.

When overhead or otherwise

When it is bright all round it will not rain; when it is bright only overhead it will.

Zinc-grey layer

A low zinc-grey layer covering the whole sky during a cold weather period is a sign of snow. If the temperature is around 32–37° F the flakes will be large and may give way to sleet; if much below 32° F they will be small.

Another useful group of weather sayings is concerned with wind and barometer changes, and these are discussed in the next chapter.

9

THE BAROMETER

IN 1843 appeared an entirely new type of instrument for measuring atmospheric pressure—the aneroid barometer. It was designed by a certain Lucien Vidie—a name now largely forgotten, even by professional meteorologists—and patented in Britain in 1844. Today the number of aneroid instruments in general use far exceeds the number of mercury-filled barometers. The former have the advantage of being light in weight, portable, and cheap to produce. Mercurial barometers, on the other hand, are heavy, difficult to transport, and costly to produce, but nevertheless give very accurate readings.

Barometers sold nowadays in the shops and stocked by jewellers in every town are almost entirely of the aneroid type

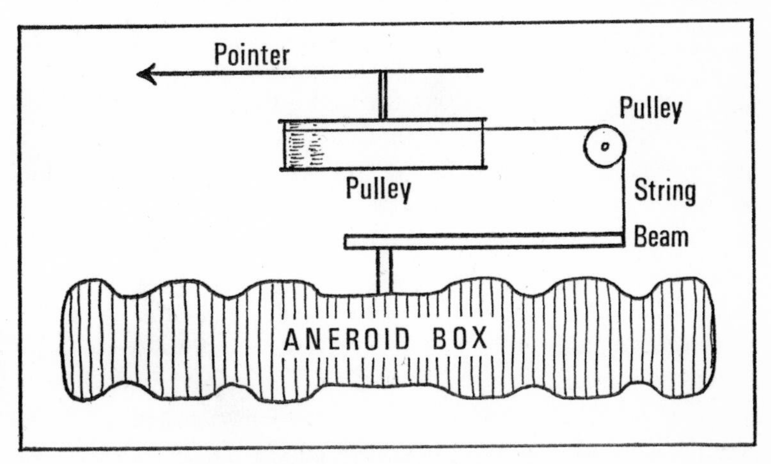

Fig. 16. Principle of the aneroid barometer.

and cost from about £5 upwards. In its simplest form the aneroid mechanism consists of a shallow capsule of thin corrugated metal which is very nearly emptied of air, the faces being kept apart by a spring. The changing weight of the atmosphere exerts pressure on this capsule, causing the metal faces to move inwards or outwards, and these movements are conveyed to the needle on the dial of the instrument through a system of levers.

Now, as in the seventeenth century, there are those who believe that the barometer is a comprehensive forecasting device and that the words placed on the instrument are to be taken quite literally: 'Very Dry' at 31 in, 'Set Fair' at 30·5 in, 'Fair' at 30 in, 'Change' at 29·5 in, 'Rain' at 29 in, 'Much Rain' at 28·5 in, and 'Stormy' at 28 in.

This scheme, although sound to a point, is too simple—simple enough to be misleading. Most people appreciate that a rising barometer points to the likelihood of better weather and that a falling barometer is a sign that raincoats and umbrellas may be necessary. What is more, with the needle on a barometer dial pointing steadily at 'Very Dry' or 'Set Fair', the chances are that the current weather situation will be exactly as stated; similarly, if the needle points to 'Stormy' or 'Much Rain'.

More often than not, however, the needle rests within the area between these two extremes and normally registers somewhere between 'Rain' (through 'Change') to 'Fair', and at these intermediate positions the weather of the present or the immediate future may prove to be quite different from what the wording of the instrument so conveniently forecasts. The point is—and this is not always made clear by instrument makers or fully understood by the public—that the words on a barometer face are intended to serve as a general guide only. What is more important for a correct interpretation of local weather is the so-called barometric tendency: whether, if the needle is steady, it has remained so for some considerable period or, if the needle has been rising or falling, how quickly this has been taking place. These movements or the lack of them need to be studied in some detail to get the best value from a barometer; not that this is in any way a difficult thing to do.

For anyone using a barometer for the first time the following 'rules' will serve as an indication of the weather to expect in a number of different circumstances. In the course of time, or as occasion demands, these may be extended or slightly modified to suit localised weather habits:

1. A prolonged slow fall *or* quick fall: rain and wind.
2. A fall of 0·5 in in less than 12 hours: sudden gale.
3. A rise of 0·5 in or more in 12 hours: remaining stormy.
4. A slow rise from low: becoming settled, but fog in winter.
5. High and steady, or high and still rising: fine and warm in summer, fog or cold in winter.
6. Slow fall from high: quick change—thunder and rain in summer; rain after fog, or snow and sleet (possibly leading to rain) after a cold dry spell, in winter.
7. Needle moving jerkily: very windy.
8. Quick see-saw motion: frequent periods of rain and gales.

A fall in barometric pressure (Rules 1 and 2) indicates the advance of a depression with its associated frontal troughs of low pressure. No two depressions or fronts are ever exactly alike, but one can be sure that once they begin to travel quickly across Britain—generally from south-west to north-east—periods of continuous rainfall and fresh to gale force winds will be strongly featured in most of the current national and regional forecasts for several days, perhaps for a week or more.

Between most rainstorms there are dry or partly dry periods with sunny spells of varying length. These are usually heralded by a quick rise of the 'glass' and may last for anything from

3 to 24 hours before the rain from the next storm. Very often there may be five or six storms in succession, the fiercest ones generally being the second, third, or fourth. This explains Rule 3 and tells why coastal fishermen have long believed that 'First rise after low foretells a stronger blow!'

A warning from the Meteorological Office of the approach of an 'intense' or 'vigorous' depression indicates that the system will be very active and will produce strong winds and probably gales over a wide area, but if the depression is described as merely 'large' the reference is to the area it covers on the weather map. It may or may not be a vigorous system, but it is worth bearing in mind that storm areas tend to reach their greatest lateral spread just before their influence begins to weaken.

Rules 7 and 8 show the qualities of more vigorous depressions. A jerky barometer needle indicates a great number of fluctuations in the atmospheric pressure within a small local area, so that not only are the gales at their height but the path of the most severe weather is probably quite close to the recording instrument. It may be several days or more before the weather really improves.

A regular see-saw motion of the barometer needle—that is, when it repeatedly covers up to half its complete range, first in one direction and then in the other—is again an indication that the main storm track is almost overhead.

Mistakes are easier to make when the barometer is comparatively high than when it is steadily falling with the needle pointing between the positions marked 'Change' and 'Stormy'. High barometric pressure is often a sign of settled weather, and it is well known that the larger anticyclones are associated with fine, often very warm, weather in spring, summer, and early autumn, but in winter (Rule 4) they bring fog to the areas lying near their centres.

The barometer needle rises very considerably before and during the coldest spells of the winter, in the same way as at the approach of summer heatwaves (Rule 5). The long rise of pressure towards the 'Very Dry' position on the dial is witness to the building up of a powerful anticyclone, and while in summer this frequently comes to Britain from either the Azores or from Russia, in winter it is more likely to approach from the east. There is a saying: 'When the wind is in the east, it's neither good for man nor beast!' and the Russian anticyclones, with their easterly winds that at times reach out to cover the whole of western Europe, are a source of drought and considerable extremes of temperature—very cold in winter and early spring, with day and night frost that becomes progressively more intense, but hot during the summer.

If a rise in barometric pressure occurs during the months of May and June and winds are easterly at the time, then there will be a marked temperature contrast as winter values suddenly give way to those of high summer, and the intermediate spring weather is then largely missed. Instances of this were more common during the 1930–60 period than nowadays, and were particularly frequent during the middle 1950s. Ground almost everywhere quickly became hard-baked.

Generally once a year—normally in the spring—the southern districts of England have a long period of rain that coincides with a steady rise of the barometer. People stare at their instruments, wondering if they are functioning correctly. But the rise in barometric pressure is in no way responsible for the rainfall; it points to a forthcoming improvement which, in this particular case, will not materialise until the slow-moving depression to the immediate south (which is causing the rainfall) collapses or becomes very weak. This may take anything up to 48 hours.

Once fine weather sets in the barometer will remain fairly steady, but it must be remembered that, at any time, a break in the weather may be heralded by a comparatively small movement of the needle towards the 'Change' position.

In summer, local thunderstorms can occur without any appreciable change in barometric pressure. In this case mainly fine weather can be predicted for the following day so long as the needle shews no subsequent fall in pressure. But a slow fall of pressure during hot weather will almost certainly lead to widespread outbreaks of thundery rain (Rule 6) within a few hours of the start of this trend. If, on the following day, the fall of pressure continues without a break, then the hot weather is unlikely to return for some considerable period and most days will be cool and changeable, with rainfall at times but also short sunny spells.

A gradual fall of pressure during the winter means a change to unsettled weather with south-westerly winds predominating. These may reach gale force near western and northern coasts, and, if conditions remain unsettled, gale force winds are likely to become more widespread at times. Any fog present at the commencement of such a fall in pressure will soon be dispersed, and if the ground has been previously frozen or snow-covered a thaw may occur. However—and this is an important proviso—a thaw may be preceded by snowfall if the first effect of the advancing south-westerly winds is to raise temperatures from below freezing to near the freezing point (32° F or 0° C) or just above it. Not until levels of above 37° F (3° C) are reached can one be certain that the rain and not snow or sleet will fall.

Many of the heaviest snowfalls of the winter are caused in this way—witness the blizzards of those memorable seasons of 1947 and 1963. There is also another point to bear in mind. A check in the barometer fall leading, as this often does, to a subsequent slow rise of pressure means that the south-westerly winds will gain no further ground and, instead, they and their associated frontal depressions will be pushed back towards mid-Atlantic by the redevelopment of a high-pressure system over the Continent. Within 48 hours easterly winds, giving ground and air frost, will be back in full strength.

Those who are interested in studying the link between weather sayings and changes in barometric pressure may have heard the rhyme:

> When the wind backs and the weather glass falls,
> Then be on your guard against gales and squalls.

The wind backs when it makes an anticlockwise change of direction, and a clockwise change of direction is known as a veer. Not all wind veers indicate fine weather, nor is a backing of wind necessarily a sign of deterioration. But a backing wind coupled with a fall in barometric pressure is normally a sign of an approaching frontal depression, and, so far as British weather is concerned, this means that the disturbance is coming from the Atlantic. If, previously, winds have been westerly they will probably back to south-west to south just ahead of the rainbelts, and if the winds were originally blowing from the south-west, they will back to south, perhaps as far as south-east. After the passage of the rainbelts there will be a veer of wind and at least a temporary rise of barometric pressure.

Remember, however:

> Should the barometer continue low when the sky becomes clear after heavy rain, expect more rain within a very short time.

Although the south-west wind brings plenty of rainy weather to Britain, many of the heaviest rains of the summer are associated with southerly winds, and the following sayings are worthy of note:

> The south wind, when gentle, is not a great collector of clouds during the summer months, but if it becomes violent, it makes the sky become cloudy and brings on thunder and rain.

Then, in terms of the barometer, here is almost identical advice:

> A fall of the barometer with a south wind is invariably followed by very damp weather and, in summer, by heavy rain.

At all times it is a fact that the barometer falls lower for high winds than for heavy rains. But when, after a succession of gales and great fluctuations of the barometer, a strong or gale-force wind coming on from the south-west causes little, if any, depression of the instrument, it is reasonable to suppose that more settled weather is near at hand.

Again, regarding improvement in the weather:

> When the barometer rises considerably, and the ground becomes dry, although the sky remains overcast, expect fair weather within a few days even if, for a time, it remains cool in those areas facing the prevailing wind.

So far as the barometer is concerned, the saying:

> Long foretold, long last;
> Short notice, soon past

fits most situations. In the case of deteriorating weather a gradual fall of the needle over a fairly prolonged period, perhaps as much as 18–24 hours initially, indicates that the approaching unsettled spell will persist for some time; as will an improvement in weather following in the wake of a fairly prolonged rise of barometric pressure. Sudden falls or sudden rises in pressure will normally produce quick reactions in the general weather situation, but these will not last long unless the falls or rises of pressure, as the case may be, are consolidated by appropriate further movements of the needle during the following 1–2 days.

When barometers are steady at 'Set Fair' or 'Very Dry' little or no wind may occur. It is customary at such times for periods of complete calm to alternate with breezy interludes when wind strength reaches about 8–12 mph. As a rule, these winds are more regular in summer than in winter; they rise in the morning and fall at sunset in inland regions of Britain, and the direction tends to be rather variable but chiefly south-easterly. In coastal areas periods of complete calm are less frequent than inland, for here the lighter breezes of fine weather blow normally from sea to shore during the daytime and back from shore to sea at night. The times that these breezes set in vary somewhat according to the time of the season and the locality but, on average, they are, roughly, as follows during the summer months: 10–10.30 am, sea breeze commences; 3–3.30 pm, sea breeze reaches maximum strength; 6–7 pm, sea breeze dies down; 7–9 pm, calm; 9–9.30 pm, land breeze commences; 3–3.30 am, land breeze reaches maximum strength; 6–7 am, land breeze dies down; and 7–10 am, calm.

Page 143: *(above)* The USA weather ship *Chain* leaving harbour for an Atlantic station; *(below)* cold front of a depression approaching from the north-west, as seen on a weather radar with range scale of 100 Km. Interval between photographs, 25 minutes.

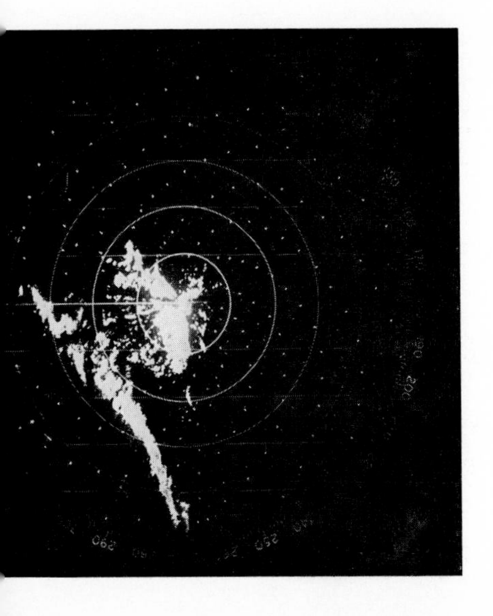

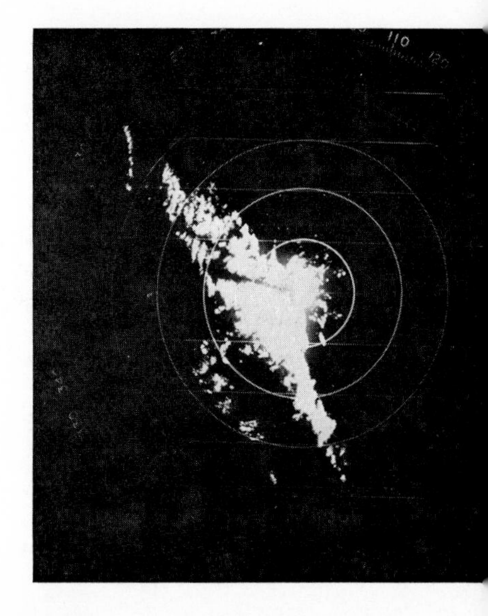

Page 144: *(above)* Thermometer screen. Maximum and minimum thermometers are mounted horizontally; at the rear are the wet and dry bulb thermometers; *(below left)* the Campbell-Stokes pattern sunshine recorder is used by weather services throughout the world; *(below right)* a modern automatic rain gauge. The rain passes into a chamber containing a float which is attached to an upright rod carrying a pen. This records the measurement on a drum-rotated chart.

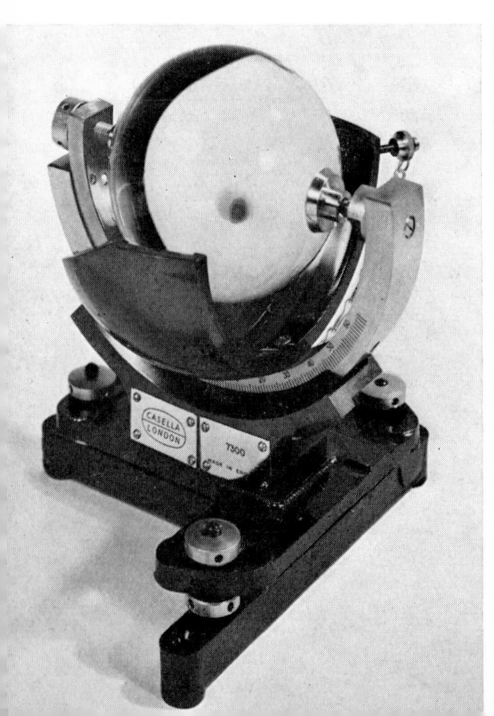

These diurnal breezes, by their very regularity, point to settled weather conditions, even without the confirmation of the barometer; if they should die out or become less regular it could well be that a gradual change in the weather is taking place and that the barometer is registering a slight fall. One thing is certain—a change to calm or near calm conditions without a high or steadily rising barometer is not to be trusted, even if the sky becomes clear of clouds. It is probably only a temporary break in an otherwise unsettled spell. And

> When there's hardly a wind—but a swell on the sea,
> For certain a drench and a gale there will be.

This is something that almost everyone who lives in a coastal area will have noted from time to time, and it occurs when the motion of the water caused by a distant storm moves faster in a particular direction than the actual weather disturbance. The barometer will begin to fall around the time that the incoming swell is first seen, if not somewhat previously, and it may be at least 12 hours before there is any marked deterioration in the weather.

Skill in linking weather trends with changes in atmospheric pressure can best be acquired by taking regular barometric readings, even when the weather seems set at 'Fair' or 'Stormy', for it is often at the moment one is off-guard that significant developments take place. It is advisable to look at the instrument at least twice daily, preferably at fixed times, and if the barometer is of the aneroid type it will be possible on each occasion to move the indicator needle to coincide with the position of the actual reading. Then, when the next reading is taken, the rise or fall of barometric pressure can be seen at a glance.

Even clearer is the continuous pressure trace made by the pen of a barograph, which is a self-recording aneroid barometer. The pen is lightly held by a spring so that it rests on the surface of a slowly revolving drum operated by clockwork. A specially graduated chart, which can be changed at weekly intervals, is fixed round the drum's exterior, and on this is marked a time scale and a scale of atmospheric pressure graduated in units of inches or, more normally, millibars, 1000 millibars equalling 29·53 in. These figures are convenient to remember, for in western Europe readings below 1000 millibars are associated mainly with changeable or unsettled weather conditions, while with readings above this limit the chances of fine weather increase, subject to the barometric tendency conforming to the rules previously outlined in this chapter.

Figs. 17 & 18. Left—sea breeze by day; right—land breeze by night.

Only very occasionally does the pressure of the atmosphere go outside the limits of 1036 millibars on the high side and 960 millibars on the low side.

Aneroid barometers and barographs require very little maintenance. They need to be set when first installed, in accordance with the makers' instructions, if reasonably accurate readings are required and it is intended to make comparisons from time to time with barometer readings at Met Office stations. The point is that barometric pressures at any given time will always fall as one ascends towards higher ground, where the weight of the atmosphere is less. Latitude and temperature will also affect the readings, and the policy of the Meteorological Office and other forecasting networks is to correct all readings to mean sea level; it would be impossible otherwise to analyse these in terms of weather changes.

A number of people who own aneroid barometers acquire the habit of tapping their instruments before taking a reading, the object being to make sure that the needle is not sticking. A very gentle tap with the end of the fingernail is all that is necessary, and to bang the dial with one's knuckle is likely to cause damage to the delicate mechanism inside.

The mercury barometers used by the Met Office, the National Physical Laboratory, and other official bodies within and outside the UK are normally of the Kew type. Mercury barometers—but built to a less exacting standard—were also made for use in the home up to late Victorian times, and some of these instruments are now collectors' pieces.

The first barometers to be made in Britain were straightforward sticks mounted on a plaque of walnut, and, later, on mahogany. But it was not long before the familiar pediments, portico tops and spiral columns were incorporated, and provision was made for the addition of a thermometer. Designs reached their peak in the eighteenth century. Inlaying with marquetry has sometimes been attempted, though this appears to have been coarse compared with that on the best clock cases and bureaux of the period.

The most interesting barometers from the collector's point of view are undoubtedly the upright portable sticks on four folding feet made by Daniel Quare, and the fine (generally Cuban) mahogany barometers of the period 1750–70. Saleroom values, however, may reach to well over £200 for such instruments, as against up to £30 for a good stick barometer of later design. One of the most expensive barometers of all was the one made for George III by Alexander Cumming, a Scottish mathematician and a Fellow of the Royal Society. It cost £2,000 to produce, and the sum of £200 a year was granted for its upkeep.

10

WEATHER ROUND THE YEAR

COMPARED with the wet and dry monsoons of India, the summer and winter rains of the African tropics, the Siberian winter snow, or the dry Mediterranean summer, the seasons in Britain appear to be almost non-existent. Or is it that they are merely hard to determine because they fail to correspond with the strict limits imposed by the calendar?

A survey made by Mr R. B. M. Levick of the British climate from 1898 to 1947 (and subsequently extended) proves that this is the case.[1] From this, however, it appears that there are five seasons each year, not four. The conclusion was reached by paying special attention to the strength and direction of the wind[2] during recurrent spells of weather which, in turn, emphasised the temperature and general weather trends. It is of course widely appreciated that British temperatures on most days and weeks of the year are largely related to the properties of the current airstream, or, when calm or near calm prevails, to those of the previous one.

Some of Levick's findings have been outlined in past issues of the Royal Meteorological Society's magazine *Weather*, and his seasons are defined as follows:

Winter, it appears, is best considered as starting about 20 November and is divided into two parts: *early winter*, from approximately 20 November until 19 January, during which time any frosts that occur seldom last more than one week and are usually briefer, and stormy periods become increasingly frequent; and secondly, *late winter and early spring*, between approximately 20 January and the end of March. Long spells of one type of weather or another are established during the late winter period, but they are of such widely differing types that sometimes this is the main period of winter, and, in other years, it has the nature of an early spring.

Spring and early summer is the next season, from approximately the beginning of April until mid-June. It contains the most changeable weather of all, for northerly winds break out from time to time and produce wintry showers, often with hail, squalls, and thunder. Then, during the early part of the summer, there is a transition from dry continental air to a moist maritime flow. On the weather maps the change is shown by the collapse of European high pressure that has spread from the east, to be replaced by Atlantic depressions moving in at intervals across the British Isles towards the Baltic or Scandinavia.

High summer, the fourth season, lasts from mid-June until about 9–10 September and is marked by the persistence of westerly and north-westerly winds, interrupted occasionally by periods of quiet anticyclonic weather.

Autumn, the final season in this classification, is from approximately 10 September until 19–20 November and generally includes at least one period of wet and stormy weather. But the early part of the season is often characterised by mild day temperatures and an occasional little summer caused by the approach of an anticyclone from the south or south-west.

Seeing that, apart from very exceptional years, each season tends to be a period of transitional rather than settled weather, the most sensible way to review the British climate's average behaviour is on a month-by-month basis. The cynic may declare that there is no such thing as average weather in Britain. Does not each year in turn, he will argue, produce very different conditions from those of the previous year?

But such argument is not well founded. Despite variations from year to year, the weather's habits are largely repetitive, as the following monthly survey (beginning with March) proves beyond all doubt.

The first part of March is normally unsettled and stormy over western Europe, particularly between Britain and the Baltic, and is associated with north to north-east airstreams. On average, the stormiest parts of this period over Britain are 1–2 and 6–9 March, and Scotland and northern England receive stronger winds than districts farther south.

By contrast, 12–19 March is normally much less stormy and often produces markedly fine conditions. In east Scotland and in many other eastern districts of Britain it is the driest period of the year. Night frost, however, is likely despite quite warm day temperatures at times. In the Home Counties and South-east England it normally remains dry until 25 March, and the period 17–25 March is the driest of the year for the London area.

There is a tradition that equinoctial gales begin on or near 21 March. In fact, they do not normally affect Britain before 24–25 March, beginning first in northern and western districts. From then until the end of the month the weather is normally very stormy, and there is a general lowering of temperature as winds blow first from the south-west or west and then veer to north-west or north. Sleet and snow over northern districts and over high ground and elsewhere are a fairly regular feature of the last few days of March, which produces storms in about 37 out of every 50 years.

The first week of April normally brings quiet uneventful weather to most districts of Britain, though it remains rather cold in northern districts, with frost at night, and this can be quite severe in enclosed valleys and over the Scottish Lowlands generally. Mid-April tends to be unsettled, particularly the period 10–15 April, which produces moderately stormy weather during approximately 38 out of every 50 years, and the peak date for storms is 14 April. Prevailing winds at this time blow generally from between west and north-west, bringing heavy showers, and these affect western districts and high ground facing west coasts in particular. Some of the showers fall as hail. This unsettled weather normally acts as a decided check in the seasonal rise of temperature, for day and night temperatures are cold. Between 10 and 13 April very cold nights are frequent in the London and Home Counties areas.

The second half of April normally opens the beginning of the thunderstorm season in Britain. The third week of April is generally quite bright and sunny, but also gives showers at times, and some of these are indeed of a thundery character. There is a risk that the moderately warm day temperatures may be suddenly countered by a day or two of colder weather, with snowstorms in some northern and eastern districts around 17–19 April.

In roughly one year in two there is a period of decidedly cool, unsettled weather between 23 and 26 April, with considerable risk of snowfall in southern England. One example was on the night of 25–26 April 1950 when a snowstorm, driven by winds of gale force, swept along the North Downs and deposited 6 in of snow. The final days of the month normally see better weather, with quickly rising day temperatures and less risk of night frost in all areas.

The month of May has two notable characteristics: first, there is a rising seasonal temperature trend, but, secondly, a marked degree of day-to-day temperature variation that makes the sequence of warm and cold spells less regular than in most other months of the year. However, there are generally two cold periods. The first occurs during 5–9 May, when there are cold winds, generally from the north-west, with showers by day and a certain amount of frost at night. A more marked cold period normally sets in around 15–19 May. Winds tend to be gusty, strong, and cold, from the north, with squally showers producing hail, sleet, and snow from time to time, and although accumulations are seldom great in southern England and over low ground generally they may be heavy locally over high ground in the north. Towards the end of this cold period winds normally decrease in strength, which leads to quite severe night frost occurring generally over Britain.

The intervening periods, namely, 1–4 and 10–14 May, produce rather nondescript conditions, with no marked temperature or general weather trend.

From 23 May onwards the weather is often fine and normally gives warm weather with less risk of temperature setbacks than early in the month. But the period has been known to produce some notable thunderstorms, at times severe and heavy enough to cause quite considerable local damage and occasional flooding; their greatest impact is in the southern and eastern counties of England.

The first week of June tends to be changeable. The first two or three days normally bring widespread thundery showers to many southern and midland districts, and occasionally farther north. Then, by around 3–5 June there is a change to cooler weather, with fresh north-westerly winds and passing showers in all areas. The drop in temperature at this time is sometimes as much as 10° F (6° C) and in some years night frost may occur in sheltered valleys.

By 8–10 June there is generally an improvement, and much of the period 8–21 June tends to be moderately fine—not always with spells of drought but with no serious interruptions apart from occasional thunderstorms or showers. From 22 June until the end of the month there are normally two to four very hot days, but this period is notable for a considerable increase in the risk of heavier type thunderstorms. At times these merge together, and heavy thundery rainfall occurs most often between 27 and 29 June, particularly in South Wales and the southern districts of England.

Between 30 June and 2 July there is generally a small drop in temperature. The first week of July is in any case a period of rather variable temperatures, with weather tending to be unsettled, certainly showery if not markedly wet, and usually some thunderstorms occur from time to time in most districts.

From 9 to 25 July temperature levels are generally higher. The thunderstorm risk continues, though the frequency of the outbreaks decreases. However, when these storms do occur, they are normally heavy, and southern England in particular can be badly affected—northern districts rather less so. Conditions generally during this middle period of July tend to be sultry and humid, with winds blowing lightly or at moderate strength from the south.

From 26 July until the end of the month it is normally cooler, with rather wet and unsettled weather affecting all districts. Moderate to fresh south-westerly winds are common and may be strong at times in the north.

The unsettled weather of late July normally continues into the first days of August, so that the first four or five days of this month tend to be appreciably colder than mid-July. After this the weather becomes more settled and much warmer during most years, and the period 8–14 August is one of high average sunshine, particularly in South-east England. Averages over long periods shows that the highest temperatures in Britain occur on 12–13 August.

After 15 August temperatures generally begin to fall and the latter half of the month is more often than not rather cool despite a 3-day warm or hot spell that may occur in either the third or the fourth weeks.

The most thundery period during the second part of August is 19–25. At first the thunderstorms tend to break out in southern districts and then move northwards to affect northern England and southern Scotland around 23–25 August. In some years there is a further general thundery period around 28–29 August, but the last two days of the month are normally settled and quite often very warm.

September often opens with another period of fine dry weather, and between the end of August and 7 September average barometric pressure rises quite distinctly in most parts of the country. Sometimes predominantly dry weather lasts until around 17 September.

But 17–25 September, which includes the period of the autumn equinox, is generally stormy, the peak being 20 September. Winds during this period tend to blow from the north-west and bring sharp falls of temperature, possibly leading to an early frost in sheltered areas protected from the prevailing winds.

The last part of September is well known as the 'old wives' summer', when nearly every country in Europe recognises the likelihood of a period of fine and warm weather. It is not so common in Britain as on the Continent, however.

October, which is the rainiest month of the year over most of England and Wales (December is the rainiest month of the year in Scotland), sets up the pattern which seems to be followed by all the winter months—stormy at the beginning and end but relatively fine in the middle. In 35 out of every 50 years there is a period of stormy weather around 5–12 October, with a peak of intensity on 8–9 October.

This windy period is associated with a succession of depressions travelling from west to east across the country and marks the beginning of London's flood risk from storm surges. Generally the wettest days in this period are 8 October in the Home Counties and south-east England and 11 October in much of Scotland. There is a steady but not a remarkable fall of temperature at this time.

In a few very unsettled years the stormy period of October leads to further storms continuing with hardly a break throughout the month, but in over 30 out of every 50 years there is a short fine period some time between 8 and 28 October, the average dates being between 16–20 October. This coincides with St Luke's Day 18 October, and is recognised in folklore as 'St Luke's little summer'. Actually, although the weather is fine and mainly dry, it is not particularly warm, the air being partly polar in origin, but sunshine brings an illusion of warmth during the day. Nights are usually cold and frosty, particularly in the Midlands and over southern England generally.

The last week of October and the first fortnight of November bring prolonged stormy weather almost every year, the average dates being 24 October to 13 November. After a storm peak around 29 October there is usually a break in the storms from 30 October until approximately 4–5 November, after which storms recur with fresh intensity. Depressions follow one another across—or to the north of—Britain, bringing abundant rain from the Atlantic. Temperatures are normally much milder than average, and frosts are rare. In fact, the frequent south-westerly winds check the autumnal fall of temperature at the end of October, but at first there is often some snow in the north. The general weather is dull and cheerless. On average the wettest day of the year in the London area is 28 October, with nearly twice the rainfall of 15 October.

Mid-November tends to be more settled, often cold, with frost at night, and night fog then develops in and near towns and is slow to clear during the daytime. The average duration of this dry period is 15–21 November. It occurs in about 33 of every 50 years, and the peak dates are 18–20 November.

The last week of November ushers in the stormy period which extends into early December, and is one of the most regular of all periods of bad weather. It normally contains two peaks of intensity: one around 25 November and the second one around 9 December. North-westerly or even northerly winds are prevalent, bringing a fall of temperature as well as increasing rainfall, but the strength of the winds prevents prolonged frosts.

It is worth recording that the worst gale experienced in southern England since the early days of recorded history occurred on 26–27 November 1703—a date corresponding with 7–8 December in the modern calendar. A deep 'secondary' depression swept rapidly across the country and blew down so many houses that the price of building materials trebled or quadrupled; the Eddystone lighthouse was destroyed and

Winstanley, its designer, was killed. Many ships were wrecked and there were record high tides in the Severn and Thames.

It was noted by Buchan, the great Victorian meteorologist (see pp 169 and 237), that 3–14 December had some justification for being classified as a 'warm period'. It is certainly a fact that in south-east England the 100 years 1841–1940 have produced somewhat warmer weather during this period than between mid-December and Christmas, but the difference is not marked. About a week of cold, relatively dry—though somewhat foggy—weather does tend to develop just before Christmas and this makes the second half of the month colder than the first in most years. The average duration of this cold dry period is 18–24 December, with a peak date around 19–21 December. Night frosts are widespread during this spell.

Around or just after 25 December begins one of the most definite of all our weather spells. It has been called the 'post-Christmas storm'. It occurs in 4 years out of 5 and its average duration is a week, from approximately 25–26 December until 1 January. The peak date is 28 December.

Structural damage done by the late December storms is generally considerable. The famous Scottish storm of 28 December 1879 destroyed the Tay Bridge, and a train carrying seventy-five people was lost in the river (see Appendix 3). The weather at this time is generally mild to begin with, but, before the end of the month, widespread snow often follows the mild weather in Scotland and northern England. It may reach East Anglia and the Midlands in some years but seldom southern and western districts to any great extent, where floods are more frequent at this time than blizzards.

January is on average the coldest month of the year in Britain. The coldest day of the year is traditionally associated with St Hilary's Day, 14 January, and in fact the three days 12–14 January do tend to bring low temperatures to London, the Home Counties, and South-east England.

But early January is normally just a continuation of the late December stormy weather, after a break of a few days after the New Year. In some years the storms continue for much of the period 5–17 January, though they are seldom so prolonged as this.

In mid-January a cold spell will often develop, with the result that there is a rapid decrease in storms. It is associated with a considerable rise of barometric pressure over central Europe and, to a lesser extent, over northern Europe. It occurs over much of Britain in 45 out of every 50 years and results in remarkable frosts during some seasons. The night of 20 January 1838 was the coldest of the century in London—4° F at Greenwich and −14° F at Beckenham, Kent. It is interesting to note that this cold period in Britain coincides with a rather regular warm period in the eastern USA known as the 'January thaw'.

Towards the end of January barometric pressure generally begins to fall again, normally quite considerably, over most of Europe, leading to stormy periods in Britain in 42 out of every 50 years. On average these periods last the nine days from 24 January to 1 February—even if there are one or two relatively fine days in the middle. Typical weather is mild, dull, and wet, but not particularly cold.

One of the most destructive of the storms of this period was that of 31 January to 1 February 1953, which brought the still well remembered north-westerly gale in which the wind reached 126 mph in the Orkneys. A mass of water was driven southwards into the North Sea by the gale, which, combining with a high spring tide, caused disastrous flooding on the east coast of England and the coast of Holland (see Appendix 3).

February on the whole tends to be a fairly quiet, dry, and moderately cold month. The term 'filldyke' applied to February is an exhortation, not a reference to its excessive rainfall. When the mild stormy period of late January and early February has ended, which may not be until around 4 February, the weather becomes gradually colder. Buchan places his first cold spell of the year during 7–14 February, and in Europe the period 7–13 February has been called the 'after-winter'. On 100-year averages at Greenwich 11 February is the month's coldest day and 11–12 February produces the coldest night. During the great frost of 1895 temperatures were recorded on 11 February of −17° F at Braemar, −11° F at Buxton, and 7° F at Greenwich. Between 9 and 17 February during that year the whole of the Thames was blocked by ice-floes, some of them 6–7 ft thick. At Oxford a coach and six was driven over the ice on the Cherwell.

Yet the cold spell does not occur with the same regularity as most of the stormy spells in Britain. The chance of it occurring in any one year is about 60 per cent, though in most years there are usually fairly considerable falls of snow in Scotland and northern parts of Britain around the middle of this month.

There is usually a short period of unsettled milder weather about 16–20 February, followed by a short return to colder conditions during 21–25 February, though this is seldom unduly severe. Finally, the last few days of the month revert to mild but mainly stormy weather.

11

WEATHER CYCLES

THE ancient Egyptians, when they discovered the regular cycles of the stars, imagined that they could apply their ideas to the phenomena of the heavens: the winds, the clouds, and the rain. Their quest was more difficult than they realised, but to a certain extent they were successful. They found an annual cycle of the seasons, which, by any modern standard, sounds incredibly simple, but in those days the exact length of the year was not known. Any successful prediction of the date when the life-giving Nile would rise in flood must have inspired awe in the common people.

From these times onwards there have been some remarkably successful weather forecasters, one of the most famous of all being Joseph. His forecast of seven fat years followed by seven lean ones laid the foundation of his fortune. It is true that, in *Genesis* 41, he cast his prophecy as an interpretation of Pharaoh's dreams, but it may nevertheless have had a scientific or partly scientific basis and been connected with a cycle in the Nile floods of about fourteen or fifteen years. In this case it is probable that people other than Joseph knew about it, and it may have been a part of the secret knowledge of the priests.

Later, when civilisation migrated northwards, the prophets found the weather very chaotic and without the pronounced seasonal rhythms of Mediterranean climes. In fact, so hopeless did the situation appear that many of the almanac-makers of medieval days resorted to sheer guesswork and had a hard time in justifying their failures.

Even at the dawn of the seventeenth century very little was

known about cycles or periodicities of weather in the north and west of Europe, and in Britain the earliest known reference is that made by Francis Bacon in his essay 'Of Vicissitudes of Things', written some time before 1625. He states:

> There is a toy, which I have heard, and I would not have it given over, but waited upon a little. They say it is observed in the Low Countries (I know not in what part), that every five and thirty years the same kind and suit of years and weathers comes about again; as great frosts, great wet, great droughts, warm winters, summers with little heat, and the like . . . it is a thing I do the rather mention, because, computing backwards, I have found some concurrence.

But it was not until 1890 that the cycle was rediscovered and placed on a scientific basis. The credit for this goes to the Austrian geographer and climatologist E. Brückner, who found evidence of it in a number of different and apparently unrelated spheres: in the variations of level of the great European rivers and of the Caspian Sea, in the frequency of cold winters in Europe, and in instrumental observations of rainfall and temperatures. After collecting further information from all parts of the world, Brückner concluded that the variations were world-wide in extent and that the overall cycle comprised spells of 10-20 cold years with excessive amounts of rainfall, alternating with spells of warm dry years of similar length. In Britain it was found that the cycle was recognisable in the widths of the annual rings of tree growth. Some years ago an examination of a 200-year-old yew in the Forest of Dean revealed that the tree had grown more rapidly in the dry intervals than in the wet ones, the well-marked growth maxima being around 1790, 1830, 1860–70, and 1900.

Then there is a 33-year periodicity identified with the Brückner cycle, which is found in the annual rings of growth of the giant sequoias of California, some of which are 3,000 years old. At Santiago, in Chile, there is a cycle of flooding that tends to repeat itself at intervals of 36 years.

The Austrian's work found immediate general acceptance. Meteorologists were convinced of its reality, and from that time onwards the average values for many of the elements making up the weather have been calculated in Britain and elsewhere for periods of 35 years, the argument being that because each period would include one rainy and one dry spell, and one cool and one warm spell, these extremes would neutralise each other to give a realistic average value.

During the present century the Brückner cycle has frequently been used to account for any unusual European weather, but what is not always realised is that it is by no means rigid and therefore foolproof. The cycle only exists when all the irregularities are swept ruthlessly to one side, and included in these are the ordinary changes in rainfall from year to year that, at the time they occur, are great enough to mask the long-term fluctuations. For example, 1921, which was the driest year since 1788, came within five years of the peak of a Brückner wet period.

Just as inconvenient is the fact that the respective periods of maximum and minimum rainfall and temperature in the Brückner cycle are calculated on an average basis; they have never occurred with precise regularity, so that to project them forward for forecasting purposes could lead to an error of as much as ten years in the timing of a change from one spell of weather to another.

A further complicating factor is the existence of various other cycles of weather that have come to light since Brückner's day. So far as Britain is concerned, the most obvious is a rainfall cycle of 51 years and 8 months which represents a slow swing from generally dry conditions at the middle and end of each of the last few centuries towards generally rainy conditions at the quarters. However, as with the Brückner cycle, it does not take into account a number of individual years which, due to short-term fluctuations, occur out of turn: for example, the wet year of 1903 occurred in a dry period and the exceptionally dry year of 1921 occurred during a wet period.

It is fortunate indeed that rainfall measurements began early in Britain. The first rain gauge was set up in 1677, and from 1727 onwards it is possible to make a continuous graph of year-by-year totals. This provides ample scope for research.

In addition to the cycle of 51 years and 8 months and the Brückner cycle of 35 years, there are weather cycles of 11 years (coinciding with the cycle of sunspots), 9·5 years, 4·7 years, 2·1 years (25 months), and 1·7 years (20 months), and these can be adjusted to form a single composite cycle. Yet even this will not account for every year of above-average or below-average rainfall. As the climatologist Dr C. E. P. Brooks so neatly put it, 'cycles are rotten reeds from which to fashion pens for writing of future weather'.[1] They appear to omit some vital unknown factor known as 'Ingredient X', which is spasmodic rather than periodic and makes weather forecasting by this method a foolhardy occupation. Sooner or later it leads to the inevitable fiasco.

Yet such is the fascination of this subject that research continues. It has been found that individual cycles are applicable in some regions of the world and not so much in others, and the sunspot cycle is a case in point. Sunspots are giant cyclonic storms in the sun's atmosphere that are seen by the telescope as small dark patches, though each covers an area larger than the whole of the earth's surface.

It has long been thought that wet years increase in relation to the increasing number of sunspots during each peak of the cycle. Alexander Buchan found that the influence could be consistently traced in the rainfall of Rothesay, in the Isle of Bute, the maximum rainfall following about a year after the maximum of sunspots. Similar recurrences were found at other stations in the western Highlands and in the Hebrides. In each year of maximum rainfall the prevailing south-westerly winds were found to be more frequent than they were at other times.

The relationship between sunspots and weather is one that continues to attract attention. Using figures supplied by Mr H. H. Lamb of the British Meteorological Office, the East African Meteorological Department in Nairobi, Kenya, has shewn that in 1906, 1917, 1927, 1937, 1947, and 1957 there was a good correlation between high sunspot numbers and high levels of Lake Victoria. But during the 1890–1920 period, and again since then, there have been times when changes in the level of the lake were out of phase with the solar cycle, so that, whatever the particular effects of the cycle happen to be, there must be some additional influence at work.

The German climatologist F. Baur has found that in central Europe there are two waves of rain increase associated with each peak level in sunspots, and a similar tendency has been noted in inland regions of the North American continent.

Mr E. N. Lawrence of the Meteorological Office gave a comprehensive survey[2] of research carried out in this field between 1945 and 1962, and, referring to Britain in particular, reported that where sunspot values have been mostly above 100 the variations in earth temperatures were nearly in phase with the sunspot cycle, but between 1890 and 1925, when the peak values were mostly below 100, the temperature and sunspot curves were in opposite phase. Lawrence has also shown that the incidence of bad fogs in the London area appears to be related to the sunspot cycle, for these occur about two years before the cycle's point of minimum solar activity. What is also relevant is that these periods coincide with the dates of maximum concentration of ozone in the upper atmosphere, and it is then that there is a relative increase in the development of cold anticyclones over Europe that favour the development of fog.

The 9·5 weather cycle is of a different type. Its cause is not known but it is well developed in many parts of the world and is related in some way to the movements of the great subtropical anticyclones. These appear to move north and south through a cycle of 19 years, giving two periods of maximum and minimum rainfall respectively to areas in middle to high latitudes, such as Britain and parts of Scandinavia.

According to Dr Brooks, who held the post of Assistant Director of the Meteorological Office and who was in charge of the climatology division, the cause of the 4·7 year weather cycle is the movement of ice by the East Greenland current. Very large amounts of ice are brought to Iceland every fourth or fifth year with almost unfailing regularity by this current, and the event is associated with strong polar high-pressure movement southwards, causing Atlantic depressions to take a more southerly track than usual. As a result, Britain tends to receive more rainfall than normal, but there are other factors that can limit the effects of this action, like the overriding influence of the shorter term rainfall cycles of 2·1 and 1·7 years respectively.

Further progress will no doubt be made in this field. For one thing, a new scheme has just been launched by the World Meteorological Organisation called the 'World Weather Watch' (see Appendix 4), whose aim is to increase the number of weather observing stations throughout the world and to improve the efficiency of existing stations. This will ensure an even better exchange of information than in the past, and computers will be widely used for the processing of data and disseminating it.

Climatic experts now look forward to the chance of widening the scope of their investigations, and much attention will be paid to statistical correlations—the study of the inter-relation of weather events occurring simultaneously or in sequence in various parts of the world. Already we have a number of important leads. Tropical cyclones, for example, tend to break out at much the same time in several widely separated parts of the world.

A further correlation has been discovered linking weather in Egypt with Britain. Apparently, a poor Nile flood in summer has often been followed in the succeeding winter and early spring by low barometric pressure and consequent stormy weather within the triangle bounded by Britain, Iceland, and Norway. A good Nile flood, on the other hand, has been associated with high-pressure within the same triangle, and with easterly winds blowing strongly across the North Sea and causing severe icing conditions on ships in this area during the winter. Another correlation of particular interest to Europeans is that average variations of temperature at Berlin in March and April have been found to be closely linked with those at Oslo in the preceding November and December.

All this confirms the truth of what was said 40 years ago by the famous meteorologist Sir Napier Shaw:

> I have come to the conclusion, that if you want to know all about the weather of Uppingham or any other place on this earth, you had better begin by looking at the weather of the world as a whole. The weather of any locality is only an incident in a very complicated turmoil of air which forms what we call the general circulation of the atmosphere, and if we want to understand any part of it we must have a working knowledge of the general system.

But this is not to deny that cause and effect relationship need always be far flung. Mr W. A. L. Marshall, who was formerly in charge of the London Weather Centre, examined the whole of the period from 1841–1949 and found that certain seasons leave their influence upon the next ones.[3] He defined his seasons as spring (March to May), summer (June to August), autumn (September to November) and winter (December to February) and came to the following interesting conclusions:

Spring

Really cold springs were followed by cool or very cool summers more frequently than by warm summers, but the hot summer of 1899 followed a cold spring when ground frost occurred until the end of May.

Mild springs were followed by approximately equal numbers of warm and cool summers.

Most dry springs were followed by summers that had less rainfall than usual. Noteworthy exceptions were the dry springs of 1852 and 1895, which were followed by wet summers.

The majority of wet springs were followed by wet summers. One of the occasional dry summers to follow a wet spring was that of 1932.

Summer

The fourteen warmest summers were followed in the main by autumns that had no extremes of mean temperature in either direction, but the hot summers of 1947 and 1949 were both followed by mild autumns.

By contrast, cool summers tended to be followed by cold autumns. None of the thirty-one coolest summers were followed by markedly mild autumns and ten were followed by very cold autumns.

Total summer rainfalls bear little relation to autumn falls of the same year. Of the thirty-one wettest summers eight led to wet autumns and six to dry ones. Of the twenty-nine driest summers four were followed by wet autumns and four by dry autumn seasons.

Autumn

Very mild autumns gave little indication of the following winter temperature. The thirteen mildest autumns were followed by mild and cold winters in roughly equal numbers. The mild autumn seasons of 1939 and 1946 were followed by very cold winters.

Really cold autumns gave a better guide to the future, and eleven of the twenty-four coldest ones were followed by cold or very cold winters and only four by mild ones.

Wet autumns gave little indication of winter rainfall, but after the twenty-two driest autumns dry winters outnumbered the wet ones by two to one.

Winter

The thirty-nine coldest winters were more often followed by cold springs than by mild ones, but two remarkable exceptions were the very mild springs of 1893 and 1945, both of which followed winters in which December and January were very cold but February mild.

The twenty mildest winters were followed by a greater number of mild springs than cold ones.

Wet winters tended to be followed by wet springs, and dry winters by approximately equal numbers of wet and dry springs.

Recent research[4] carried out by individual members of the Royal Meteorological Society has established a connection between April temperatures and the general character of the weather during the following summers. The records used were those for Newark, Notts, and they clearly showed that none of the five warm Aprils that occurred during the period examined (1949–66) were followed by cool summers, and the three outstandingly warm, dry summers of 1949, 1955, and 1959 were all preceded by warm Aprils. Another investigation, using the Manchester records, proved that warm Julys are more likely than not to be followed by warm Augusts, though such a temperature persistence from one month to another had often previously been noted in Britain and in many parts of Europe and Asia.

Yet another line of research revealed an apparent connection between the onset of easterly winds in May in the upper atmosphere in east Scotland and the resulting ground level conditions during the following summer in the south-east of England.

As for rainfall, it has been shown that, for a period of up to a week, the longer a wet spell lasts, producing successive days of rain, the greater are the chances that it will continue; this applies to the south-east of England, and similar results have been found in other regions. For the north of Britain the chances of successive rainy days increase between the first and third days of a particular wet spell, and then decrease for the next two days. However, if the spell does in fact continue after the fifth day, the risk of further rainy days steadily increases until after the ninth day.[5]

Similarly, for a fine spell lasting just over a week, the chances of a rainy day following a fine day decrease day by day between the first and the seventh or eighth day (according to the area).

It is often affirmed by countrymen that Nature always 'pays her debt', and for this reason any instance of fine warm weather occurring out of season is taken to be a bad sign and to indicate that, before long, we shall have to endure cold or unsettled weather again. 'A January spring is worth naething', runs an old Scottish proverb, and another more sweeping one informs us:

> If January calends be summerly gay
> It will be winterly weather till the calends of May.

Or, again,

> Warm February, bad hay crop;
> Cold February, good hay crop;

and 'A windy March and a rainy April make a beautiful May', runs another countryman's saying.

Yet another declares:

> A cold April
> The barn will fill.

These sayings, and others like them, were born more of a desire to see an orderly cycle of seasonal weather behaviour than from the fact of its existence, for fine warm summers have followed unusually fine and warm springs, as in 1959, and, just as frequently, cold disappointing springs have been followed by equally disappointing cool summers. Nature balances her books very erratically and is careless about her arrangements of hot summers, cold winters, floods, and droughts and the other weather phenomena. Very seldom do we have to 'pay' next year for this year's eccentricities; more often than not we have to face the same kind of excesses, pleasant or otherwise, for several years in a row.

There are times, one must admit, when it is difficult to separate weather fact from weather fantasy. For example, if an analysis of the good and bad summers is made for the last eighty years, it will be found that, with the exception of 1903, which was extremely wet, the good summers (very warm, sunny, and dry for most of the time) occurred during years with odd numbers: namely, 1899, 1911, 1921, 1933, 1947, 1949, and 1959. The bad summers (very cool, dull, and wet for most of the time) occurred, with one exception of 1903, during years with even numbers: namely, 1888, 1890, 1912, 1920, 1954, and 1956. It was calculated by Sir Graham Sutton that the odds against this arrangement being due purely to chance is about 50 to 1, although there is no recognisable pattern in the sequences themselves.[6]

Sometimes the relationship between a particular season and the following one can be detected in advance by amateur meteorologists keeping a close watch on the published Atlantic weather charts or those devised by the BBC. If, from early or mid-May onwards, the Greenland anticyclone becomes very large, leading to the extension from it of a high-pressure ridge first in one direction and then in another, then this system is likely to be a regular feature of the weather charts during the early part of the summer as well. Furthermore, in terms of western European weather, the development of high pressure to the north will mean that Atlantic depressions take a more southerly track than normal, and a number of these will find their way via Biscay into the Mediterranean. This will result in at least a wetter-than-average start to the summer in the south of Britain, but in Scotland—and also to some extent in parts of northern England—the weather will be drier, possibly quite sunny, due to the reasonably close proximity of the northern high-pressure system.

Reference has already been made to Alexander Buchan who, in 1867, discovered what are now known as his 'hot and cold periods', when the current weather tends to exaggerate the effects of the calendar seasons or runs counter to them. To use Buchan's own phrase, these are 'annually recurring interruptions in the seasonal rise and fall of mean temperature'. The dates of the cold periods are 7–14 February, which claims to include what are normally the coldest nights of the winter; 11–14 April, coinciding with the continental 'Blackthorn winter'; 9–14 May, which is also the 'festival of the ice saints' in many parts of Europe; 29 June–4 July; and 6–13 November. The dates of the warm periods are 12–15 July, which claims to include what are normally the hottest days of the summer; 12–15 August; and 3–14 December. Today these periods are frequently quoted as though they apply to the whole of Britain, which is not so.

A quarter of a century after Buchan's death his name became a household word in Britain. It happened like this. In 1927 and 1928 Lord Desborough's Bill for fixing Easter was under debate in Parliament. This measure sought to enact that instead of ranging over five weeks of March and April as it had done for centuries—and is still doing—Easter should always be celebrated on the Sunday immediately following the second Saturday in April—that is, on one date or other between 9 April and 15 April.

But it was objected in the House of Commons that if Easter Sunday were to be confined to the proposed limits the associated public holiday would, in most years, clash with Buchan's second 'cold period'. In the end the objection was overruled and Lord Desborough's Bill reached the Statute Book. There it stands, but with the Act remaining in suspense pending sanction from each of the various Christian churches.

Reference in Parliament to the Buchan periods at once aroused popular interest in them. In 1929 it happened that nearly all the cold and warm periods arrived close to schedule, and the public's interest grew very considerably. It was even suggested that Buchan should be canonised and made the patron saint of British weather.

So Buchan achieved posthumous fame beyond his wildest dreams. Yet he himself was the first to declare that his cold and warm spells were not regular in their occurrence. It must also be emphasised that Buchan's research was confined to the records of Scottish weather stations: Sandwick, in the Orkneys, representing the northern and more insular parts of the country; Callton-Mor, in Argyllshire; Glasgow, representing the western districts; Milne-Graden, in Berwickshire, representing the eastern and drier areas; and, finally, Braemar, Aberdeenshire, for the Highlands.

Without doubt, Buchan's name lives on because he was one of the greatest ever researchers into weather cycles. His knowledge of weather and climate, and the use that he made of this, proved to be very extensive. When he was secretary of the Scottish Meteorological Society in the early 1860s, his studies on the effects of weather on crops were so successful that he was able to use rainfall and temperature statistics to foretell the price of cereal and root crops several months in advance. No one since has been able to repeat the achievement.

12

WEATHER AND WARFARE

THE history of all ages affords examples of battles or campaigns lost or won as a result of the weather's overriding influence. At least, these are claims that have been made. It is the task now of modern research to discover to what extent these claims bear close examination. Were some of them just excuses made by defeated armies in order to cover up man-made mistakes?

Results show, beyond doubt, that a number of invasions and battles have succeeded despite adverse weather conditions, but we are then left with the interesting question of whether or not a particular victory would have had results of greater consequence had the weather been more favourable. Certainly the Roman Occupation of Britain would have been more effective in its early stages if Caesar had been favoured with good invasion weather, and the Viking raids would have been far less telling and also less frequent had the prevailing climate been unfavourable at the time. One can also point to the stormy weather of the Middle Ages—when good all-round climatic conditions in Europe were restricted to the extreme south of the continent—to explain why military activity was severely restricted near British coasts.

After this period reports of the weather's influence on the course of our history become more numerous, and, at that time, because we had ourselves acquired the taste for taking the offensive, it is not surprising that the scenes of natural—some thought supernatural—intervention on the part of the elements shift to foreign soils and, even more frequently, to the seas around our coasts.

The battle of Crécy presents a good field for enquiry. To just what extent did the weather really interfere with the course of events? The relevant weather report comes from Sir John Froissart's *Chronicles* (translation by Lord Berners):

> Also the same season there fell a great rain and a clipse (flash of lightning) with a terrible thunder and, before the rain, there came flying . . . a great number of crows for fear of the tempest coming. Then anon the air began to wax clear and the sun to shine fair and bright, the which was right in the Frenchmen's eyes and on the Englishmen's backs.

It is safe, no doubt, to rule out the possibility of the English troops having received an accurate weather forecast, with the ability to act upon it. Here, however, was a case of a dramatic change of weather giving benefit to one side, and as it happened to the superior of the two armies.

So, too, at Agincourt. There had been heavy autumn rains, and the ground was sodden. The French horses stuck in the mire and were shot down with their riders. The French men-at-arms who tried to attack on foot could barely drag their feet through the mud. But it was the French who had chosen the place of battle; it was the French who made the attack; it was the French who sat their horses all night in the rain and were weary men at dawn. Can very much blame for their defeat be fairly laid upon the weather?

On one occasion when a storm did in fact have a direct effect upon military operations, its action was to deny victory to both sides. This was in the early spring of 1360, when Edward III was besieging Paris, and the French were offering to negotiate for peace. Edward chose to fight on, but by mid-April it was apparent that the city was too strong to be captured; then, at a critical moment, there came a devastating thunderstorm, with great hailstones, that caused very heavy casualties. Contemporary estimates put the English losses from this cruel storm at 1,000 men and 6,000 horses. Even if these losses were exaggerated there is no doubt that the storm, by destroying Edward's baggage train, crippled his army to such an extent as to convince him that negotiation was, after all, the only course. The treaty of Bretigny—admittedly an ill-observed one—was a direct result of the storm.

On the whole, the effects of adverse—in most cases, stormy—weather on the course of British history appear to be more clearly defined in battles on land than at sea; for the number of unpredictable factors at sea is always bound to be greater.

On one famous occasion it was not a storm but fog that directed the course of conflict, for on 14 April 1471, at Barnet, a dense fog so confused the opposing forces that the Lancastrians mistook their own troops for Yorkists, fired upon them, and in the resulting confusion were defeated. Rainfall, with or without hail, wind, and various other forms of inclement weather, was, however, more potent on most battlefields than the mere lack of visibility. The two great military disasters that befell Scotland—Flodden in September 1513 and Culloden in April 1746—both occurred in very wet weather, which had a considerable effect on the operations.

At sea, whatever the complications and apparent 'freak' effects of adverse weather, it would nevertheless be difficult to find a single campaign in which the plans of one or other admiral were not upset at some point. During the Persian and Punic wars, losses at sea were very great as a result of sudden storms; but a nation's resources counted for more in those days than its seamanship. Conditions changed when capital ships ceased to be galleys, and the sail replaced the oar; the effort needed to replace a lost ship—or, worse, a number of ships—was then greatly increased—witness the destruction by gale of the Spanish Armada in 1588, when, having been defeated by Drake, it tried to escape round the Scottish coasts.

But because the Spaniards recovered as quickly as they did and, with their new fleet, captured the *Revenge* at the Azores three years later, it seemed that the gale that destroyed the Armada had just delayed the course of history. As it happened, Philip prepared a second Armada nine years after the first.

This time our tactical intelligence was faulty. It had been assumed that October was too late for an invasion (correctly, as it turned out), and because of this Lord Howard of Effingham had taken the English fleet on a plundering cruise to the Canaries. Philip was determined. His second Armada set out but encountered a fierce westerly gale, and peasants of the Biscay coast reaped the same harvest of wreckage as the Scots and Irish had reaped before. Yet the war with England dragged on for a further twelve years—and with the Netherlands for very much longer.

Rainstorms and thunderstorms played a full part in the drama of the Civil War, and while there is no evidence of a single storm creating victory or defeat on its own, one might well speculate on what course the war would have taken had the Royalist camp at Marston Moor taken as much notice of the weather as Cromwell did. John Buchan reminds us that at 7 o'clock, when the rain had gone and the sky had cleared, the Parliament forces decided on battle. 'There was time enough, and light enough. . . .'

Buchan further asserts that 'but for the victory at Marston Moor Parliament would have gone down, its armies would have melted away, Leven and his Scots would have re-crossed the Tweed, and Charles in six months would have been back in Whitehall. . . .'

Later that year, on 26 October 1644, the weather favoured the Royalist cause at Newbury. Manchester's support for Cromwell arrived too late in the planned encircling of the well fortified ground that protected the royal troops. Before the battle finished the sun had set, clouds came up, and in the night the king moved off, unmolested, towards Oxford. Charles was, however, fighting defensively; to escape with his life at that time, with or without the aid of weather, was all he could hope for.

During the first war between England and the Dutch a storm intervened just in time to prevent the occurrence of a major sea battle. It was in the summer of 1652 that the rival fleets of Blake and Van Tromp had sailed north, each for its own reason, and found themselves in each other's presence off the Shetlands. But before battle could be joined a severe gale swept down upon both fleets. By luck the English had the shelter of the land and suffered little; but the Dutch, who were exposed to the full force of the wind and were on a lee shore, ceased for the time being to exist as a fleet. Many of their ships were wrecked, and the rest were scattered and disabled; and it remained for Van Tromp to make his way home as best he could.

This was a major trial, but it was in no way decisive. Within two months the Dutch had their fleet at sea again and were able to gain the victory in a sequence of battles and skirmishes. One could say, at the most, of the Shetland storm that it weakened the Dutch navy at a crucial time, and to that extent prejudiced its chances of success in the battles that were to come.

There was something, however, to be said for venturing out to sea—usually during a storm or period of contrary wind—when the opposing side least expected it. It meant taking a great risk, that is certain, but it was a carefully calculated risk that William of Orange and his soldiers took when, on 1 November 1688, they decided to set sail for England. Visibility was poor and the wind was strong, from the east, so William passed unnoticed through the Straits of Dover while the English fleet of James II was kept in harbour. When the news did break through, William was sailing briskly down the Channel. His good fortune nearly came to an end when his first attempt to land at Torbay was unsuccessful and he was driven towards Plymouth, a strongly royalist port which would have given him a most unfriendly reception. But then, when the whole cause seemed to be lost, the wind, after first slackening, shifted half round the compass and took William straight back to Torbay. He landed at Brixham, having been at sea for five days, and in course of time established for himself his unique position as joint ruler. Even today, the wind that carried

William safely to this country can be claimed as the most influential gale in our history.

From the many eyewitness accounts and from the evidence of the widespread damage caused, the storm of November 1703 caused as much damage as a fully fledged tropical hurricane. London and Bristol might just as well have been bombed with high explosives, so great was the havoc. Trees, houses, and steeples toppled. Defoe counted 17,000 trees down in Kent alone. At sea whole fleets were cast away, though, by curious chance, the great ships were all spared. Yet in 1744 a normal type of autumn gale was sufficient to sink the largest and newest ship in the British navy, the *Victory*, of 100 guns. It took down with her Admiral Sir John Balchen and 1,100 officers and men.

This was indeed a tragedy, but it was no reason for being defeatist; quite the reverse. Within a few years of this event the great ships were not only keeping at sea during the winter season but were doing so without suffering disproportionate losses. How far the credit for this change lay in better construction and equipment, and how far it should be sought in the improved seamanship of the officers and crews, has not been (and perhaps cannot now be) exactly decided.

In any event, the British nation seems to have derived a certain amount of gain from many of its storms, especially in the sea engagements of 1588 and 1652 with the Spanish and Dutch fleets respectively.

During the eighteenth century we seem to have been less fortunate in this respect, particularly in 1744, when a gale denied this country what appeared to be an almost certain victory. The French, having just begun their war with Britain, were bent on invasion. An army and transports were collected at Dunkirk, and a fleet was sent up Channel to cover its passage. On 23 February the French vessels anchored at Dungeness, but on the following day Sir John Norris came in sight with a superior fleet. At a council of war the invaders decided that it was time to be gone, and that they must weigh as soon as the tide served; the English, without doubt, would have weighed at the same time and could hardly have failed to intercept the enemy. But at this juncture a severe gale arose from the north-east. The force of the wind was such that fighting was out of the question, and the quarter from which it was blowing was fair for the retreat of the French. They ran before it, returned safely to Brest, and, as a result, the invasion of England had to be postponed.

The advantage, however, went solely to the French, for the invasion, as so often happens to postponed projects, never took place. If it had taken place the French fleet would have been defeated, and the war would not have dragged on unsatisfactorily for several years for lack of an English victory at sea.

Other sea battles have been prevented by inopportune gales: for example, there was the bad weather of 1692 which prevented the arrival of the French Mediterranean fleet in time to help with the projected attack on England; the stormy March of 1708 which delayed another French expedition so that it only reached the Firth of Forth a few hours before Admiral Byng appeared with a fleet twice as strong; and the equally stormy March eleven years later that brought to nothing one more invasion from Spain. On another occasion, in 1778, the fleets of Howe and d'Estraing were separated off Rhode Island by a gale that drove both sides into port to refit. As soon as any battle was joined, a calm or a shift of wind was just as likely to prove an important factor as the arising of a storm.

Gales, however, have frequently been present after a battle to deny the winning side the fruits of victory. A hurricane sank Rodney's lame ships and prizes in 1702; a strong wind made it impossible to keep the damaged *Warrior* afloat after Jutland. To take the former case, Rodney's prizes included the Compte de Grasse's flagship *Ville de Paris* which, had she been brought into port, would no doubt have served as a model for the improvement of English first-rates. Valuable time was lost, for no other first-rate was captured from the enemy until 1793, when the *Commerce de Marseille* was brought from Toulon. Her design helped to give us the larger and more powerful ships which began to join the fleet after the war with France was renewed in 1803, but, had it not been for the loss of the *Ville de Paris*, we might have had such ships when the war broke out in 1793.

Three years later the weather came to our aid. Ireland just then was singing an exciting new song, 'The French are in the Bay'. And so they were, but Bantry Bay at Christmastime is a tempestuous place, and soon it was evident that the ships carrying the cash and stores was missing. Just as it had been agreed to land the army and hope for the best, the ships were enveloped in a sudden squall, and this stopped all thought of disembarking. The French soon left the bay.

Gales that blew continuously between 22 and 25 October 1805, after the battle of Trafalgar, are among the most infamous in our history and caused the loss of no less than fourteen line-of-battle ships that had been captured by the English, as well as of two, one French and one Spanish, that had not. It had been Nelson's plan to anchor the fleet as near as possible to the scene of battle, but Admiral Collingwood decided on the opposite course when he took over command.

One storm in our history resulted in the saving of many lives that would otherwise have been lost. This was in 1789, when the *Adventure* was driven ashore near Tynemouth. She was fast only 1,000 ft from the shore but a fearful sea was running, and many of the local inhabitants watched helplessly from a headland while, below, their relatives were drowning.

A few days later the people of South Shields held a meeting where feeling ran high and passionate speeches were made. Someone suggested that a special kind of boat might be designed for saving lives at sea, and it was decided to offer a prize for the best design. Henry Greathead won it with his famous *Original*—the first specially designed lifeboat in the world and the prototype of many installed in other countries.

Week of 14–20 June, inclusive: a broad storm area is moving in from the Atlantic; several weather fronts associated with this storm will produce intermittent rain early in the week and general continuous rain through midweek. Gradual improvement will occur during the latter part of the week. Mobility of artillery and heavy vehicles will deteriorate rapidly, becoming poor by midweek.

24-hour forecast for period ending noon 18 June: General warm front rains will develop during the night of 17 June over a wide belt, continuing through most of the night and well into 18 June. Cloudy conditions will continue throughout the night of the 18th. Mobility for artillery and heavy vehicles will be greatly impaired, as will the ability of foot soldiers to manœuvre. Artillery fire control will be impaired by poor visibility.

If Napoleon had had a meteorological staff equipped to deal with the weather in the modern sense, these are in essence the forecasts he might have studied before he dictated battle orders on the evening of 17 June 1815 to two generals who scribbled them on paper supported on straw pallets on their knees.

But the French leader could have had no advance weather information worthy of the name, and if any had been offered to him he probably would have thought it sorcery or an 'old wives' tale'. He had always regarded short-term changes of weather as a disagreeable but trivial nuisance, brushing the topic aside impatiently. His early Italian and Austrian campaigns included many battles fought both against his human foes and against hostile weather.

Now, in ninety days after returning from Elba, he had put together a formidable and well equipped army, and it was aimed directly at the English under Wellington and the Prussians under Blücher, who were just south of Brussels but separated by a few miles. Four nations had covenanted to invade France and take Napoleon, like so many badgers digging out a fox, but the Austrian and Russian armies were still far away. Napoleon's plan was to smash the English and Prussian armies before their allies could arrive. His whole art and concept of war, which had carried him to the mastery of Europe, was to smash his enemies one by one. He was the master of surprise, of the rapid manœuvre, and the smashing blow at the weak spot. He liked to 'aim his artillery like a pistol' now here, now there, but his tactics required above all a dry field and a firm footing for rapid movement, and in the supreme battle of his life these were denied him.

The British lay before Waterloo the whole of the night of 17–18 June beside camp fires. The French troops across the valley lay in the darkness, and suffered under the pelting rain from midnight until a dawn that was only a partial lightening under the streaming clouds.

Napoleon realised now that he had to overpower the British before Blücher could rally from his recent beating at Ligny and join forces with them. But he also appreciated the handicaps that the deep mud of the soggy grainfields imposed upon rapid manœuvre. He had no meteorologist whom he could ask: 'How soon is this going to clear up?' So he decided to wait awhile before attacking, in the hope that the sun would come out and dry the soil to some extent. But the sun did not come out, and the ground did not dry up. It was 11.30 am before Napoleon accepted the hazard and ordered the attack, and in those four wasted hours he lost France and the world.

For the delay gave the indomitable Blücher, firmly marching the twelve miles that separated him from contact with the French army, urging on his exhausted men as they pulled the wagons and artillery out of slough after slough, just enough time to come up in the late afternoon and deliver the fatal blow at Napoleon's reeling troops.

'A few drops of water . . . an unseasonable cloud crossing the sky, sufficed for the overthrow of a world,' says Hugo. Napoleon gambled, and Nature beat him—though some allege that the British have always given Wellington more credit than the weather. Napoleon might well have known better, one might think, since it was climate that toppled him in 1812 when he marched to the east. The Russians simply retired before him, allowing the Russian winter to close upon the Grand Army.

When that terrible grip relaxed, only a few thousand emaciated 'walking corpses' staggered back to France. Again, in 1941, Hitler's armies walked right to the gates of Moscow and Russia seemed almost within their grasp—then the terrible premature winter closed down and froze the ill-clad supermen by hundreds and thousands. But here the comparison between the two campaigns must end, for Hitler had good meteorologists yet refused to allow retreat. Napoleon, when he laid his plans, had asked Laplace to find out when the winter really began to set in over central Russia. Using such statistics as were available (admittedly they were not over-abundant in those days), Laplace replied 'January'. The Emperor acted accordingly, a severe cold spell came in December, and the campaign was lost.

Waterloo, though, represents what was up to that date history's supreme example of weather changing human destiny in war. It put paid to Napoleon's career and to France's ambition to dominate Europe.

From then until 1940 no invader's plans got beyond paper. The First World War, however, showed us a new way in which the winds might influence a battle. Gas had become a weapon of war and gas went where the wind blew it. The prevailing winds of the Flanders battlefields were west and south-west, which was a natural advantage to the Allies, but the importance of a sudden shift of wind was enormous and unforeseeable. Then, in 1918, the last-moment preparations for the German spring offensive coincided with a warm dry spell, and on 21 March, the day it was launched, the attack was first made under cover of thick fog that made defensive measures very difficult. It was said that the Germans had advance knowledge of the fine weather, which lasted for several days, whereas we did not.

So we come to 1940 when western Europe had one of the sunniest springs on record. Week after week the exquisite peaceful sunshine ripened crops before their time, and the beautiful June that followed gave perfect weather for hay and the promise of a great harvest for those that could attend to the reaping. 'Hitler's weather' held for what seemed an almost indefinite period of time.

In all wars weather has contributed to both victories and defeats on land and sea. It is totally without concern for human affairs, and is a third force willing to fight for that commander or nation that is best able to make use of it. In World War II the German pocket battleships made brilliant tactical use of Atlantic mists and areas of low cloud, as did the German army in their final desperate thrust in the Battle of the Bulge under cover of weather that kept the Allied air strength blinded and grounded. The Japanese invaded the Aleutian Islands, which lie in one of the world's great storm-producing regions, under cover of continuous bad weather that kept American scout-planes on the ground. The only warning that reached the American garrisons there was based on a correct long-range estimate of the weather moving on the Aleutians from the Pacific, together with the significant fact that the Japanese were preserving radio silence. The Japanese later evacuated Kiska under cover of a similar series of storms, and so successfully that when the Americans attacked they struck an empty island.

The great evacuation at Dunkirk was made possible by the bravery of the British fighter-pilots in pouncing on the German planes that were seeking to bomb and strafe the helpless huddled troops; but the British air cover might not have held out for ever. As it happened, a kindly pea-soup fog crept over the Channel and blanketed it for several days. Under its shelter, and with the sea absolutely calm, the British brought off 300,000 men, impressing into the service every kind of craft that would float.

On the other hand, the American fleet under Admiral Halsey, which had withdrawn from Luzon in the Philippines to refuel after three days of strikes against that island, was caught on 17 and 18 December 1944 in the grip of one of the greatest of all typhoons. There were twenty carriers, eight battleships and numerous small craft, apart from the twenty-four tankers they were to meet for fuelling. The Navy meteorologists could not agree on the position of the storm centre, partly because of the meagre weather reports available from the vast stretches of the Pacific. Many ships wallowed helplessly in the 'dangerous half' of the typhoon, where the whirl of the rapidly circulating winds is reinforced by the speed of the system's forward movement as a whole, so that the overall wind force is greatly increased. A wind speed of 145 mph was registered at one time. The fleet lost 790 men and several ships, and, save at Pearl Harbor, suffered total damage worse than any the Japanese ever inflicted.

A naval court of inquiry found 'large errors in the predicted path of the typhoon', said Hanson W. Baldwin of the *New York Times*, and naval commanders were instructed to see that all their officers mastered the 'law of storms', which is simply a century-old formulation of the way cyclones and hurricanes move as a whole. Operating on a far larger stage, Admiral Halsey apparently had little more trustworthy weather information than Lord Nelson might have had a century and a half earlier. The Pacific typhoon seems to have been well nigh as complete a surprise as was the Japanese attack on Pearl Harbor and the sudden swoop of Japanese aircraft from the dense cloud mass of an equatorial rainstorm to destroy the British capital ships *Prince of Wales* and *Repulse*.

Yet correct weather information supplied to the Allied commanders played a great part in shortening the war, and modern meteorological methods served the Allies particularly well on three decisive occasions: during the great air strikes that preceded the Normandy invasion; on D-Day itself; and, finally, for the launching of the decisive European campaign in the spring of 1945.

The weather prologue to D-Day goes back to the early career of Dr Millikan, who was chief of the science and research division of the Signal Corps of the US Army in World War I and who acted as a government consultant both before and during World War II. He always believed in a weather science that would adequately meet demands of all kinds, warfare included. Between the wars he created a meteorological unit for serving the specialised needs of industry, and in the early 1930s students were assigned to it by the US Signal Corps and Air Corps. Many students were also sent there later by the US Weather Bureau and the Navy, and the department gained recognition as the nation's outstanding unit for training military personnel in ways of applying modern meteorological methods to the problems of war.

About a year before Pearl Harbor General H. H. Arnold, then Deputy Chief of Staff, US Army Air Forces, and later Commanding General of the US Army Air Forces throughout World War II, visited Dr Millikan's laboratory. By that time its young meteorologists had completed their first full-scale studies of certain rhythmic variations in the prevailing westerly winds of the northern hemisphere and had developed from these studies a basic system of long-range forecasting. They were confidently and successfully analysing atmospheric situations thousands of miles away, and forecasting weather much further in advance than others dared to do. Even more significantly, they were forecasting for areas from which the local weather reports, previously considered indispensable, had been blacked out for security purposes.

One such area was Newfoundland. A company that harvested Christmas trees asked the laboratory during the summer of 1940 to inform it as far as possible in advance when Newfoundland might expect at least a week of good weather during the late autumn, so that crews might go into the forests to cut and haul out the trees. The technicians in California watched the succession of cyclones and anticyclones floating north-eastwards across the United States, and compared the forming and dissolving weather patterns with those of previous years. At last they saw a large high-pressure area, with its accompanying promise of clear weather, on the way towards Newfoundland.

They sent a telegram telling the company when to go ahead.

It happened later that a reply had just come in from the Newfoundlanders, expressing thanks and appreciation for a forecast that had worked out perfectly, as General Arnold walked into the laboratory. The General read it, and asked if Newfoundland itself was not blacked out as to all local weather reports. He was told it was, and his keen mind instantly leaped ahead to the military possibilities of such a method of long-range weather analysis. It could be used, for example, to predict many days in advance the weather that would prevail over Germany and the huge area of Europe then controlled by her. And, indeed, it was.

The weather requirements for D-Day and its preceding and succeeding operations were briefly as follows:

> For several days before, winds must not raise heavy continuing swell in the English Channel; this for the Navy's benefit. Actual wind strength to be less than 14 mph in the Channel, and not above 19 mph immediately outside.
>
> For aviation engaged in the transport of troops and materials, a ceiling of at least 2,500 ft and visibility of three miles. For heavy bombers, a sky not more than half covered with clouds below 5,000 ft, and ceiling not lower than 11,000 ft. For medium and light bombers, the ceiling could be no lower than 4,500 ft, visibility not less than three miles over target. Fighter pilots needed at least 1,000 ft between clouds and ground surface.
>
> For paratroop landings, the wind over the target area should not be more than 20 mph at most; for glider landings not over 35. For both, moonlight, twilight or dawn should illuminate the ground.

This is no means the most complex situation that can be imagined in these days of three-dimensional warfare. Had the carriers and the fleet been involved as strongly as in the Pacific, there would have been added a whole set of extra conditions based upon the swell of the sea, the speed of the winds combined with the carrier-speed to permit take-off, upon the relative speeds when each carrier received back her planes, upon cloud-cover, defence against aircraft, and many other such considerations. The function of England as an aircraft carrier, anchored sturdily in the sea, was a very valuable consideration in the success of D-Day.

Past weather records showed that the odds were best in June for the desired conditions. The high command in the late winter fixed upon a date around the first of June for D-Day. Late in April the commanders-in-chief tentatively set the date for 4 June, with 5 and 6 June as alternatives—and with the next 'possible' period as 17 to 21 June.

The final D-Day forecast team was made up of six persons —two each from the British Air Ministry and the British Admiralty, and two from the US Strategic Air Force headquarters. The British, using the traditional short-range method of forecasting, could see no weather coming up that would justify the risk of committing the great expedition to the stormy Channel crossing. The Americans, using their newly developed long-range forecasting methods, saw 'possible' weather. They also foresaw continued worsening weather for some weeks, if the early June opportunity had to be missed.

General Eisenhower probably never knew how bitter was the division and argument within the forecasting team as to whether the next few days would provide any weather opportunity at all for the invasion. But, at last, with 3,000,000 allied troops poised for the invasion, a decisive weather forecast was issued at 9.30 pm on 4 June. Those present at Supreme Allied Headquarters were General Eisenhower, Admiral Ramsey, Air Chief Marshals Tedder and Leigh Mallory, General (as he then was) Montgomery, and Eisenhower's chief of staff, Major-General Walter Bedell Smith. All day long a steady gale had lashed the English Channel, and now the three senior meteorologists entered the conference room, led by Group Captain Stagg.

Eisenhower knew that he had to choose between 5, 6, and 7 June for the invasion. The rain and wind had already ruled out

4 June, and the convoys which had sailed almost to the invasion coast had been called back at the last moment, to the confusion and the dismay of all present. If the invasion was to be put off by even one more day the refuelling of ships which had already been a day at sea could cause the whole delicately balanced operation to be postponed for weeks, with calamitous results.

Stagg opened the briefing in hushed silence. He referred to the meteorological picture of the previous twenty-four hours and then quietly said: 'Gentlemen, there have been some rapid and unexpected developments in the situation.' A new weather front had been spotted, which, he declared, would move up the Channel in a few hours and cause a gradual clearing over the assault area. These improved conditions would last throughout the next day and continue up to the morning of the 6 June. After this the weather would deteriorate again.

A single day of fair weather was less than what was needed, but, on the strength of this one forecast, Eisenhower ordered the great movement to start. In *Crusade in Europe* he tells how a wave of relief and optimism swept all commands, irked and chafing from the days of sitting in bitter uncertainty because of the weather. A forecast ruling the weather 'impossible' might well have delayed for a year the ending of the war.

A final piece of irony is that Major Lettau, the chief German meteorologist, and his staff agreed that the weather following 4 June would be much too bad to permit an invasion attempt, because of new storms moving in from the North Atlantic. As a result, the German high command had relaxed: many officers were on leave, and many troops were on manœuvres. Later in the month, between 17 and 22 June, a prolonged storm smashed up many of the Allied artificial harbours and set back the tempo of the invasion by weeks, but Eisenhower's decision had been correct, for all that. He made this comment: 'Thanks, and thank the gods of war that we went when we did'.

The new longer-range techniques in weather forecasting rendered one other specific major service during the European war. This was to assess the likelihood, in 1945, of the usual spring floods on the Rhine. History showed that the Rhine is likely to flood at any time until about 1 May, because of rains and melting snow high on its watershed. Should the spring offensive of 1945 be started before that date, floods could have wiped out supply lines thrown across the wide river, and stranded Allied forward units helplessly on the other side without support. The weather men were asked what was the likelihood of flood. A long-range forecast of precipitation possibilities showed that a very dry March was extremely likely throughout all of northern Europe in 1945, thus lessening the danger of April floods in the Rhine almost to the vanishing-point.

The analysis was given to General Eisenhower's intelligence section in January 1945. It placed odds on crossing the Rhine early, without threat from flood, at about 92 per cent. On this assurance, the attack was launched against eastern France and western Germany in February 1945, and the Allies crossed the Rhine on 23 March.

Thus the allied offensive swept through western Germany at least seventy-five days ahead of the timetable that would have been imposed by waiting until 1 May. This early start undoubtedly brought the war to an end sooner—and also, had the western Allies waited on the flood hazard, the Russians might well have swept across all Germany to control the Ruhr, the industrial heart of Europe.

What of the future? It can be expressed very simply. The weatherman, seldom in the front line of history, must continue to give vital counsel behind the screen of military secrecy, particularly as his role may well extend to modifying the weather as well as analysing and forecasting it.

13

WEATHER CONTROL

IN 1965 a small team of scientists from the University of the North, at Antofagasta in Chile, set themselves a most unusual task—to provide water in the rainless Atacama Desert, short of carrying it across the mountains by muleback. It seemed hopeless until they noticed that thick mists from the Pacific drove inland across the desert during the afternoons, and as they did so drops of water condensed on to the leaves of the giant cacti. This gave the leader the idea that perhaps more of the fog might be condensed into actual water by using the appropriate device, which was a frame threaded with nylon mesh resembling a harp that could be set up on poles across narrow chasms in the path of the incoming fog.

The experiment met with immediate success. With the arrival of every fog water collected in droplets on the nylon mesh and, on falling to the bottom of the frame, was collected into a trough that led to a storage tank. One 4 ft deep frame strung vertically with 1 mm nylon thread at eight threads to the inch collected water at the rate of 4 gallons per square yd on heavily foggy days, averaging 220 gallons per square yd per year.

Two further lines of enquiry were then suggested. In the first place, if this was the amount of water collected from a single small frame, the quantity could be increased at will by using extra or larger frames. The fogs occurred regularly, so that the source of water supply was virtually constant.

The second question concerned the action of the frames on the fog itself. Presumably, by extracting moisture in this way, the density of the fog would decrease—at least in the area where the frames were situated. In this case, would the effect be sufficiently great to make a material improvement to, say, visibility over motorways or other roads where fog was a menace?

Late in 1965 the challenge was taken up by Mr W. R. Bellis, the Director of Research at the New Jersey Department of Transportation. First he conducted laboratory experiments using rotating frames—'fog brooms'—inside a specially constructed chamber 30 ft long, 20 ft wide, and 10 ft high. This was filled with sufficient fog to reduce the visibility to 5 ft or less. With 10 brooms in operation, the fog cleared in 4–5 minutes, and each broom collected approximately 6 cc of liquid water per minute. The same fog required about 30 minutes to clear by itself without any mechanical aid from the fog brooms.

On 30 June 1967 the system became operational. An automatically operating revolving fog-broom installation was set up at Parkway Avenue, Trenton, New Jersey. It was designed so that it was turned on by a photo-electric device when fog occurred above a certain density and turned off when visibility improved to the required level. The installation comprised twenty individual rotating brooms evenly spaced along 320 ft of the Avenue.

On 8 September the first seasonal fog occurred. The installation turned itself on automatically at 5.43 am and turned off 37 minutes later. The recorded data showed a partial clearing of the fog at the installation. The next fog to affect the area occurred on 17 October. On this occasion the value of the clearing process was more difficult to assess because the fog began to lift of its own accord at the same time as the brooms were in action.

The New Jersey authorities say that each new fog teaches them more about the equipment and that there is much they have learnt already. It is likely that this method will become widely used in the course of time.[1]

The latest use of nylon-broom equipment is on the Rock of Gibraltar, where water supply has always been a long standing problem. Mr E. A. J. Canessa, a chartered civil engineer, who is a resident, had been struck by the anomaly of the town having to economise on fresh water whilst there was so much moisture suspended directly above it. Every year untrapped water shows itself in the form of the well known Levanter cloud, the banner-shaped formation that forms when easterly winds strike the Rock with any force.

Mr Canessa decided to set up a simple experimental nylon screen 4 ft high by 2 ft 6 in wide on Signal Hill on the east side of the Rock, just above the main water catchment area. Provision was made for tilting the screen into the wind at the most favourable angle, and the 1968 results (the first year of operation) were encouraging. The frame collected small but meaningful amounts of water from passing formations during 6 months out of the 10 in which tests were made, and, during October, when there was a prolonged heavy Levanter, nearly 10 gallons of water were collected.

If further tests are successful a suitable installation should be capable of collecting worthwhile quantities of water, and the apparatus would have the distinct advantage of being able to operate at virtually no cost outside the initial capital outlay.

These experiments show that we need no longer talk in terms of 'if and when' man will be able to control the weather. This rather negative attitude is met quite frequently, perhaps because troubles caused by extremes of weather, particularly excess of rainfall or the lack of it, seem to be perpetual. In Britain, except in very dry years like 1959, most districts suffer from too much and too frequent rainfall during the summer months, yet only very occasionally do crops in south-eastern districts receive the desired amount of rain during their respective growing seasons. In fact, there is a belt of approximately 50,000 square miles nearly every spring from east Devon to Birmingham and Edinburgh, and from Edinburgh to Kent that suffers from drought, and there is also the problem of replenishing our underground water reserves which, due to a steadily increasing consumption, are in danger of becoming exhausted.

Of course, the unequal distribution of rainfall, and the consequent often catastrophic effects, is a problem known to practically every part of the world. More than ever the current need is to even out Nature's benefits for the general good of all mankind, but experiments designed to increase or limit rainfall have not been wholly satisfactory. This is the view of many British meteorologists and of the Meteorological Office as a whole.

The history of rainmaking goes back many years and crude 'methods' are still widely practised today in Australia, Africa, and parts of America, as they were many centuries ago. 'I have watched an aboriginal "wise man" in Arnhem Land squirting mouthfuls of water in all directions, as an optimistic hint to the sender of rains', wrote Mr J. B. Sidgwick.[2] The question one has to answer today is, simply, are the modern methods so scientific and effective that we can afford to pour scorn on the devices used by primitive tribes?

To answer this question it is necessary to review the progress made since the period that started before the turn of the present century. In 1891 the United States Congress approved the expenditure of 9,000 dollars because, at the time, it was claimed that rain could be 'jarred' from the clouds by sound vibration. The experiment went ahead using balloons filled with an explosive mixture of oxygen and hydrogen. High explosive shells were also pumped into the clouds, but it was all to no avail. Then there were some other ideas that proved equally useless, most of them showing a neglect of practical feasibility and complete indifference to the welfare of any who got in the way; these included the scattering of sulphuric acid and sand into the clouds from aircraft. Successive failures did not stop other bands of optimists from repeating experiments of this kind for some years afterwards: for instance, rockets were fired into the sky from Hampstead Heath during the drought of 1921.

The first important experiments in rainmaking were made in 1946 by Dr Irving Langmuir and Dr Vincent Shaefer of the American General Electric Company. Dr Shaefer discovered that by introducing pellets of solid carbon dioxide (popularly known as 'dry ice') into the supercooled layers of potential rainclouds (see p 12), countless millions of tiny ice crystals formed, all of which were capable of quick growth by feeding on the surrounding moisture and could fall to the ground in the form of rain or snow, depending on the temperatures in the lower levels. However, to stimulate the precipitation of rainfall for crops and water supplies by dropping dry ice from aircraft was likely to be an expensive and tricky business.

The problem appeared to be overcome when, during the following year, the American physicist Bernard Vonnegut obtained similar results by using silver iodide crystals. In the form of a smoke it was possible to feed them into the clouds at an estimated rate of 30 quadrillion per minute by using generators that burned the chemical at a very high temperature.

It was easy to see how rainmaking operations could become economic, for silver iodide is a reasonably cheap chemical and the operation of ground generators is comparatively inexpensive. However, it has been alleged that the very simplicity of this method led to its commercial exploitation long before its effectiveness had been thoroughly tested, which unfortunately was—and still is—a difficult thing to do. In the first place, one can seldom be sure, when operating at ground level, that the silver iodide particles always reach the correct layers of the cloud; they might all be blown away by turbulent winds. Also, this particular chemical is affected by sunlight and in certain conditions it can lose its vital powers after about an hour or so of exposure. Some of these drawbacks can, however, be overcome if operations take place to windward of high mountains, for here the rising particles can be quickly taken up into the very cold levels of the cloud where their action is likely to be immediately effective.

During the 1950s the commercial rainmakers in the USA and

Canada built up a vast business with ranchers, farmers, and water authorities, and contracts were obtained for carrying out operations abroad in Mexico, Spain, France, and Scandinavia. In Norway operations were intended to switch rainfall from lowland to mountainous areas by seeding suitable cloudbelts over the watersheds. Here the hoped-for extra precipitation would fall as snow, increasing the winter snowpack and eventually supplying the hydroelectric power plants with the additional water required. In Israel experiments in rainmaking went hand in hand with irrigation and desert reclamation projects, so as to obtain cumulative and lasting benefits.

The contracting companies engaged in rainmaking operations applied results tests by comparing the normal rainfall for each season for the general region in which they were operating with the rainfall actually measured in the operational target areas. Many successes were claimed, yet a large number of independent scientists without commercial ties have asserted all along that this method is not valid. They argue that rainfall is the most variable of all the quantitative measurements in meteorology and that it is not unusual to record heavy rainfall in one area and little or none a few miles away. Therefore, how can one tell after an apparently successful seeding experiment that it was not Nature, after all, who provided the rainfall in the right place without any help from man. This being so, one should not claim successes except after a very prolonged series of trials in a particular area.

Nevertheless, the general opinion in the USA during the early 1950s and subsequently was that the case for rainmaking was proved beyond all doubt, even if at times overstated. A Congressional Advisory Committee on weather control was formed and various States, beginning with Wyoming, passed laws establishing ownership of all clouds that travelled over their respective boundaries. There was a very real fear that rainfall could be stolen by one State from another by the drying out of clouds during their course of travel.

Nor was this the only problem. Claims amounting to some

2,000,000 dollars were filed in 1950 by New York citizens who claimed that artificially induced rainfall had caused harm to them in various ways. Then, in the autumn of that year, when New York city reservoirs were nearly empty after a long dry summer, a Harvard University meteorologist was called upon to plan and supervise a programme of aerial cloud seeding. This was successful to the point of embarrassment, for 15 billion gallons of water fell into the New York reservoirs while neighbouring States still complained of drought.

Among promising rainmaking experiments of recent date are those that have been carried out in many parts of south and east Australia by the radiophysics division of the Commonwealth Scientific and Industrial Research Organisation under the direction of Dr E. G. Bowen (no relation to the author of this book). It was officially stated that these experiments have led to an increase of rainfall over large areas of as much as 25 per cent and that there is no doubt that this is additional rainfall, not simply a redistribution. The closely controlled experimental work has shewn that, with silver iodide, *cumulus* type clouds having the right droplet size can be made to precipitate up to 1,000,000 tons of water, which is equivalent to 1 in over 15 square miles.

Rainmaking operations should be capable of bringing economic benefits to comparatively small areas within a single season amounting to many millions of pounds, and the greatest benefits are likely to be experienced in marginal rainfall areas such as the drier parts of Australia and of North and South Africa where the maximum possible amount of precipitation must be wrestled from the appropriate cloud formations during short rainy seasons or periods. Parts of South Africa, for example, suffered from the effects of drought for much of 1966–7, particularly the Transvaal, the Orange Free State, and the Cape.

In Britain the first large-scale rainmaking operations were begun by a team of Meteorological Office scientists over Salisbury Plain in 1955. Immediately there were strong protests from farmers in the area, but repeated official assurances, 'We cannot make rain fall from a blue sky!' were more than confirmed by the long periods of drought and the scorching heat of that year. In fact, suitable clouds for seeding, which need to be at least 18,000 ft thick, were so scarce that in the end operations had to be postponed until the autumn.

They were recommenced on 5 October, and on that day the rains were heavy and, for some, disastrous. But Mr B. C. V. Oddie, the chief operator, denied making more than just a small proportion of the rainfall.

The trouble in Britain is that, at any one time, there are likely to be fewer people desiring rainfall than wishing for it, and it has even been suggested that we should deal with natural excesses of rainfall by seeding Atlantic rain clouds some considerable distance from our islands, so that, by the time they reach British coasts, they are partly dehydrated and unlikely to create any flooding risks. Such a scheme, however, is quite impractical, for in generally stormy and very moist weather conditions the effects of an operation of this kind would be unnoticed. Cloud seeding is thought to precipitate, at the most, one per cent of a cloud's water content in the immediate vicinity of operations, and this amount is soon made good by the airstream's moisture reserves.

Apart from rainmaking, a branch of weather modification that claims many current successes in several parts of the world is that of hail control. After two years of experiments in Kenya, designed to protect tea plantations, it was found that after firing high-explosive rocket shells into the hail-producing areas of storm clouds, damage was consistently less in the protected areas than outside them, and in the USSR, where rockets are used to seed the clouds with silver iodide,[3] Acadamician E. K. Fedorov has stated that hail damage in the protected areas was three to five times less than in the unprotected regions. The problem of hail of course, is a very old one, and it is not confined to Africa and Asia. North America suffers very acutely from hail.

Another notable area for the production of frequent and severe hailstorms is that of southern and central Europe, particularly the mountainous regions, and hail is of course capable of destroying whole vineyards within a few minutes. In the Dark Ages primitive tribes in these hail-ridden areas would shoot at advancing storms with arrows. Then, from the sixteenth century onwards, guns were used, though for much of the eighteenth and nineteenth centuries the method was declared illegal by the current rulers. But in 1896, in Steiermark in Austria, after twenty years of very severe hailstorms, a certain burgomaster named Albert Stiger decided that something had to be done. He decided to reintroduce the old-fashioned hail cannon, and within a few months he had set up thirty-six 'stations' to protect the valleys of Steiermark. His own town of Windisch-Feistritz was defended by no less than seven stations on the surrounding hills, all positioned at different heights.

The first year of operations saw an immediate reduction in hail damage in the region, and Stiger's methods were widely copied. Yet doubts lingered in some quarters as to the effectiveness of the operations; moreover, casualties from accidental explosions were considerable. There were eleven deaths and sixty serious injuries during the year 1900 alone. In 1902 the Austrian government held an international conference in Graz to investigate the matter in detail, and it was decided that the success of the hail cannon, though unlikely, was not disproved, and that further tests should therefore be made.

Two test areas were selected and operated respectively by the Austrian and Italian governments, and for the next two years both used barrages far greater than any that Stiger had created. Up to 20,000 shots were fired at times into a single storm, and these were capable of reaching a height of about 6,000 ft.

The result of these experiments came as a victory for the sceptics, for hail damage was considerable during these two years. Since then firing at hailstorms has continued spasmodically until the recent rekindling of interest. The fact is that though a few spectacular 'successes' do not vindicate the method, neither can it be said to be totally ineffective without scientific proof or carefully controlled observations over a long period. One trouble is that it is easy to theorise on what happens to selected areas of a hail cloud that is treated with high explosives or with chemicals, but it is quite another to obtain really convincing observations of a scientific nature. This being the case, there must always be the feeling that while the cost of operations is certain, the benefit is unknown and therefore dubious. The same argument can be applied to other known devices for suppressing hailstorms, which include the de-electrification of the clouds by using a massive battery of masts fitted with special conductors and treating the clouds with ions produced from radio-active material.

One of the dangers of many weather modification attempts, even if they appear to be purely local, is that unforeseen consequences may arise. This is particularly the case with atmospheric treatments, using chemicals, where the precise cause-and-effect relationship is not known. Cumulative effects are also possible, and one process designed to achieve a limited objective may initiate something quite different. Mr Walter Orr Roberts, Director of the National Center for Atmospheric Research in Colorado, recently stressed the importance of low-energy trigger actions in major changes in weather patterns. His view is that man is already influencing his environment to an alarming degree and that it is now essential to opt for an international regulation of all actions than can affect the atmosphere. The problem could well become even more pressing than the control of atomic weapons.

An example of how man may be interfering with the weather without realising it has been given by another scientist, Professor G. J. F. MacDonald of the University of California. He alleges that 400 supersonic aircraft making four flights per day on inter-continental routes at heights of around

60,000 ft, would introduce 150,000 tons of water a day, as well as enormous quantities of carbon dioxide, into the atmosphere. These would eventually lead to a lowering of the amount of heat radiation from the earth's surface, and though the average temperature increase over the world as a whole would be less than 0·5° F, this could lead to other types of climatic change, including a possible increase in storm activity in certain regions.

It is therefore significant that the US National Academy of Sciences has asked for an increased level of research into all problems affecting weather modifications during the 1970s. Man, says the Academy, 'can and does interfere with the atmosphere in a number of different ways. His ability to produce deliberate beneficial changes is still very limited and uncertain, but it is no longer either economically or politically trivial'.[4] This statement is in very direct contrast to the views held between the two world wars. W. H. Humphreys, a well known meteorologist in his day, declared in 1926 that rainmaking would remain impracticable, for 'man would have to put enough energy into a cloud to wrestle it down directly by main strength, so to speak', and 'it would cost the equivalent in energy of 36 million horsepower exerted continuously for a week to compel Nature to provide one inch of rain over so small an area as ten square miles'.

Of course, there are other ways of modifying the effects of weather apart from making experiments in the atmosphere. Farmers and fruitgrowers exert some form of climatic control, both in long-term planning and in day-to-day operations.[5] The effects of heavy frost for example, may often be reduced by removing sections of fence or hedge to allow freer circulation of air. In addition, crops can be protected from frost in a number of ways. The traditional way is to space a battery of oil burners round an orchard and they will raise the general temperature level by up to 4° F. Smoke-producing 'smudge pots' have also been used but are not too efficient. Another device, which has been tested in many countries, involves the use of fan heaters; and another, which has not so far been widely adopted in Britain, uses rotating blades on poles to bring down warm air from a higher level and mix it with the chilly air that accumulates nearer to the ground on potentially frosty nights. But it is extremely difficult to keep warm air near the surface level.

Frost can, however, be minimised by the use of the correct cultivation and husbandry methods, particularly for crops of low growth, such as potatoes and strawberries. Because it is warmer by night over bare soil than over grass, and warmer over most compact soil than over loose dry soil, the best conditions for frost-free cultivation are those that can maintain well weeded, consolidated moist soil, and early irrigation may be desirable. Observations have shown that a field of new potatoes that has just been irrigated can escape damage during a late spring frost, whereas a neighbouring one that has not been irrigated is likely to suffer considerable damage.

Irrigation is one of the factors that can increase crop yields, so much has been known since the earliest times, but only in recent years was it discovered how to estimate the changing irrigation need of a crop from week to week. A team of scientists from the Met Office Agricultural Branch, headed by Mr L. P. Smith, perfected a special irrigation technique for use in Britain and elsewhere. First, one must estimate the amount of moisture that has been transpired by a crop, using current weekly rainfall and temperatures in relation to previously prepared tables; then, from this, one can estimate the amount of irrigation required so that the crop can continue to grow without suffering any damage. To look merely at the state of the ground, whether it is wet or dry, may not be sufficient.[6]

Investigations have also been carried out to help improve the climate on farms by the erection of various types of 'shelter belts'. These are normally semi-open barriers which filter the wind but do not stop it abruptly and can be made of lath fence, wire-open-mesh netting, or even bales of straw. Dense windbrakes such as walls or very thick hedges suffer from the disadvantage of producing damaging downdraughts on the lee side of their barriers. Tests showed that most early crops need protection from the cold dry east and north-east winds of spring, and though cauliflowers provide an exception to this rule and do not profit from any form of shelter, beans, early potatoes, lettuce, and anemone crops give better yields behind laths or wire-netting. The straw bales, which help new potatoes, have an adverse effect on beans. Whatever the size of the field to be protected, maximum benefit for crops can only be obtained if sufficient crosswind length is provided, which, in practice, amounts to somewhat more than twenty times the crops' height.

Intentionally or not, the long-term climatic patterns of large areas of country are being gradually changed in many parts of the world. In a large and well heated city the air temperature in winter may be raised by 4° F or even more by the warmth escaping from factories, offices, and domestic buildings, even allowing for the fact that this is being constantly dissipated. Local weather effects can also be produced by the replacement of one ground surface by another. For example, tests carried out in Arizona have shewn that a layer of asphalt can increase maximum afternoon temperatures 0·5 in below ground by up to 20° F. This will lead to increased cloud production, as *cumulus* can be formed by currents of air rising from heated ground, and, as a result there may be a greater risk of showers on some days. The same rising currents of air provide a useful lift for gliders, for they are in fact strong thermals.

To what extent a global modification of climate can be achieved by altering the landscape is a branch of the science that is still in its infancy, and arguments regarding its feasibility and the probable benefits and snags of suggested projects will no doubt continue for a long time. One very intriguing proposition is the Russian scheme to dam the Bering Strait in order to change the north polar climate. The idea is to pump cold Arctic water into the Pacific, and, when this process is set in motion, it should make it easier for the North Atlantic's warm ocean current, the Gulf Stream Drift, to melt the ice as it flows past Iceland and Greenland towards the Barents Sea.

It is well known that ice reflects a large proportion of the sun's radiation back into space, so that to remove it is likely to make the Arctic warmer. But there could also be a countering effect caused by the resulting northward movement of the Atlantic storm areas. This would increase cloudiness and precipitation, and therefore the snowfall over the Arctic, which, in turn, would induce a steady build-up of the Canadian and Greenland ice caps and make the polar climate more severe in these regions.

The possibilities in this field are very great, but international co-operation is necessary, and it is unthinkable that any project of this kind should be started without a thorough investigation into all its aspects. Costs would vary from case to case but could be small in relation to the benefits achieved if the chain of cause and effect is attacked at its weakest link.

In the meantime those who make a study of weather and general climatic modification find that its diversity is one of its great appeals. For example, while the New Jersey State Transportation authorities were experimenting with their fog-broom equipment, Norwegian weathermakers were recently very successful in their attempts to turn fog into snow at Oslo airport. By spraying the fog from aircraft with finely powdered 'dry ice' they caused the fog to thicken for some twenty minutes and then to precipitate to the ground as snow. After repeating the process several times a relatively clear corridor was swept along the approach path, and, as a result of this treatment, the airport was kept open for two days during the first serious fog of the winter.

Another new method of clearing fog from runways, called 'Turboclair', has been developed by a French company aided by the Aeroport de Paris. This uses the heat and kinetic energy of jet engines to vaporise fog droplets. The flow is directed along the ground, mixes with the surrounding air, and lifts the fog near the end of the runway, and practical tests have shown the effectiveness of the technique. Up to fifteen jet engines may be needed to clear a runway; they must be located in specially constructed pits to avoid causing any obstruction, and fitted with deflector funnels so as not to create turbulence.

The cost of running such an installation is not excessive in view of the advantages gained. During World War II the then newly discovered fog dispersal method, known as FIDO, employed oil burners massed along the runways, but costs were prohibitive and the method was not adapted for regular peace-time use. But it was used on one occasion during the great fog of November 1948 to enable a Viking airliner to take off from Blackbushe, Surrey for West Africa with urgently needed currency.

Outside the boundaries of scientific weather control one does hear from time to time of people who claim to be able to influence the weather without using apparatus of any kind. There is a man living today in Dorset who concentrates his thoughts on *cumulus* clouds in order to dissolve them, but these clouds are in any case constantly building up, varying in shape and size, and dissolving of their own accord; on warm days during spring, summer, and early autumn they begin to disappear in inland regions during the late afternoon or early evening, so that, in this particular sphere of activity, all things are indeed possible.

But why stop at the clouds? In past eras the control of winds could be a remunerative occupation for those who claimed the necessary powers. Marco Polo wrote of the inhabitants of Socotra, in the Indian Ocean, that they can 'cause the sea to become calm, and at their will can raise tempests, occasion shipwrecks and produce many other extra-ordinary effects'. The Scandinavians, however, appear to have been the most prolific wind-raisers of all and it seems that the traditional art must have spread southwards. Sir Walter Scott tells of one Bessie Miller who lived at Stromness, in the Orkneys, in 1814:

> Her fee was extremely moderate, being exactly sixpence, for which, as she explained herself, she boiled the kettle and gave the bark the advantage of her prayers, for she disclaimed all unlawful arts. The wind thus petitioned for was sure, she said, to arise, though occasionally the mariner had to wait some time for it.

Even during the middle of the last century there were times when Scandinavians and sorcery were regarded in the same light, and captains of vessels burdened with persistent head-winds were keen to discover the presence of a Russian or a Finn among the crew who could be used as a scapegoat and placed in irons until the wind eventually changed.

14

CLIMATE, WEATHER, AND MAN

IN a world of many climates, one is tempted to ask, where is the human being most efficient? In a survey made between the wars a prominent researcher, Dr William Cramer, declared that the climate of England came nearest to the ideal. Also 'approaching the ideal' were the climates of Japan, New Zealand, and the south-eastern corner of Australia. On the American continent Dr Cramer considered the most efficient climates to be within the fairly broad belt between southern Canada and the eastern and central USA and also along a narrow stretch of the Pacific Coast from British Columbia to California.

But it has been found that the areas of the world which are best for human efficiency are not those that are the most pleasant or the most comfortable. Nor are they necessarily the best from the health point of view. Climate, which affects our rate of growth, speed of development, fertility of mind and body, and the amount of energy available for thought and action, has a dominating effect on our health. Yet to say which is the healthiest climate in the world for any particular individual is no simple matter. In warm climates people are more susceptible to infectious diseases, but in varied climates like those of Britain and North America the prevailing cool and stormy conditions for much of the year provide the human body with too much stimulation and lead to frequent breakdowns in its machinery.

One can of course argue that ailments caused directly by climate are few; sunstroke and frostbite are examples. What happens is that climate—and with it, seasonal and weather changes, where these are relevant—plays an important role as a predisposing agent in the cause of many diseases. How this works out in practice is an interesting study.

Consider, first, the middle and higher latitudes of the various continents. The widest areas of these are situated in the so-called temperate zones of the world and are found chiefly in the northern hemisphere, because it is here that the land areas are greatest. Here the nations are leaders in world affairs. Our American and European neighbours, and ourselves as well, have achieved great things through our initiative: we boast vast towns, motorways, skyscrapers, dams, bridges, and many other impressive monuments. We have traded with tropical countries for long periods, which has added to our material wealth. And now we reach for the planets. But we have paid a price, and still do, for this kind of activity. Whilst slow-maturing people in hot climates have difficulty in losing body heat, our own ready ability to get rid of it drives us on to activity that causes stress, and this appears to increase all the time as a result of growing urbanisation and the complexities of modern living. It is an alarming health menace because, by or after middle age, the body seems unable to endure this kind of strain with equanimity. In short, the cumulative failure over the years to give the body proper rest and relaxation tends to leave the system prematurely old, and it is in the parts of the body most concerned with the supply of energy that evidence of breakdown is most frequent.

A prolonged forced draught in the human firebox means that the oxygen must be carried from the lungs to the body cells at an unusually high rate, overworking the heart and blood vessels. The troubles resulting from this include excessive hardening of the arteries; hypertension, more familiarly known as high blood pressure; coronary disease; and apoplexy or stroke, which is the rupture of a hardened artery in the brain. These diseases are all killers which, while they tend to spare the weak and infirm, strike down energetic people, but in the tropics coronary attacks and heart failure are uncommon except in advanced old age.

These facts may make us doubt that the changeable climates of America and western Europe are the best regions of the world in which to live. In addition, because the red cells of the blood are the real carriers of oxygen to the system, it is in these same climatic regions where heart failure is most common that one finds the most frequent cases of exhaustion of the bone marrow that produces those red cells. The condition, known as pernicious anaemia, is again a disease that is very rare in the tropics.

Another prevalent disease in our own stimulating latitudes is diabetes, which represents a breakdown in the body's ability to transform the food eaten into the special kind of sugar that can be burned in the cells.

The American scientist Professor Clarence A. Mills, who has carried out much painstaking research on this subject,[1] relates that he spent his childhood and early youth on a farm in Indiana where climate acts as a dynamic force on life and the frequent weather changes have a disturbing influence on the daily functioning of the body. At the age of seventeen he moved to South Dakota, where the climate is even more invigorating. He had numerous colds and sore throats during the Indiana winters and, as a result, the strain left some mark on the heart muscle, which by middle age had become sensitive to stress or too much stimulation. The result was that winter conditions led to a rise in blood pressure and an inability to relax, while summer heat brought welcome relief from this kind of stress, with the blood pressure returning to its normal level.

In the autumn of 1926 Professor Mills left America to work on the staff of the Union Medical College at Peking, and, during the following summer, for the first time in his life he experienced several months of complete lethargy. He discovered that the climate of northern China, despite its high latitude, had a totally different quality from that of the energising regions of North America. For the summer heat in Peking was not counterbalanced by a stormy winter season. There was also the additional factor that the Chinese winter here was not interrupted by sudden warm spells, nor was the summer disturbed by days of refreshing coolness. This, he considered, explained why there was less evidence of storm-caused illness there than in his native region of America. Subsequent research proved that Americans and British people find that their blood pressure falls during a long stay in Peking, often by 30 per cent during the first year.

Professor Mills relates a further case of a journey from the stimulating American climate to the tropical heat of the Philippines. He left New York on a January day during a blizzard, on board a slow freighter, and as he entered the soothing Caribbean warmth and sailed past the coasts of Central America and Mexico, he could feel his inner machine gradually slowing down. Soon he was able to sit and relax for hours of the day without having the feeling that he should be doing something or planning some new activity. After being in the Philippines for a comparatively short time his blood pressure was down below normal but rose to a very high level during the following winter in North America.

The Philippines, it should be noted, are not completely ideal for health, for the northern islands lie in the westward half of the track of the oriental typhoons en route for south China, and this seasonal storminess is reflected in the fact that the Philippine capital Manila has more hypertension cases than would normally be experienced in the tropics, but nevertheless, fewer than in most North American cities of comparable size. Similarly, in the West Indies, where autumn brings its annual crop of hurricanes, hypertension is relatively greater than in tropical areas not affected by such storms.

Every summer brings a moderate fall in the blood pressure of many people, but it takes a long heatwave to exert any pronounced effect. However, when summers are unusually warm and prolonged in the northern hemisphere, as they were in Britain during 1955 and 1959, the metabolic rate of many of us will change and our tissue fires and blood pressure will subside to tropical levels. For this reason a number of people with diabetes or restricted working capacity of the heart can allow themselves a somewhat greater freedom of activity in established summer warmth without ill effect; in any case it is true that everyone, no matter what his state of health, uses up less energy in doing a particular job of work in summer than in winter.

In hot countries the afternoon siesta is an established custom, and those who have migrated from energising climates are in need of it, for they are prone—women in particular—to nervous exhaustion if they try to continue with their normal active life in the tropical heat. But, curiously, it is in the world's most energising areas, such as our own here in Britain, that the habit of rest and relaxation is most needed, particularly for those who are entering middle life and showing signs of wear from previous decades of activity.

Climatic effects upon human beings are often accentuated by their eating habits and the effects of drugs. In the tropics B vitamins are particularly necessary, for these help the body to work at the greatest possible efficiency by speeding up the burning of food and helping to counteract the sluggish effects of continued heat, but the meats consumed are often low in these dietary factors and are made lower by prolonged cooking. The net effect is to emphasise the lethargic pace that is typical of these regions. In higher latitudes an opposite but equally unfortunate set of circumstances exists. Here many people who are driven by climate to a restless activity in their daily lives turn frequently to drugs and stimulants, which place a further burden on their straining bodies.

One of the greatest stimulants of all is caffeine, which is found in coffee, tea, and cola products and serves to speed up the burning of food in the tissues and increase their functional activities. The body becomes more alive, the mental processes keener, and the heart works harder. Most tea and coffee drinkers admit that they drink these beverages for the pick-up that they give, but continual heavy use of the caffeine stimulant brings on exhaustion that calls for ever increasing doses; so, while it is fine to use it for brief periods of need, its over-use to whip up flagging spirits is harmful. It tempts the human to perform more daily work than his system can stand and may also cut down his sleep, for the effect of caffeine does not wear off entirely within a few hours. But in tropical countries the use of caffeine normally presents little danger, since the response of native people to its energising action is comparatively low.

As with caffeine, so with alcohol. In warm regions, where human energy is low, alcohol produces only a mild excitement phase, and this quickly passes. However, excessive amounts taken by those who, on moving to the tropics from stimulating climates, suffer from boredom caused by inactivity, can create serious problems; these people find that they cannot burn up the alcohol in their bodies as rapidly as they did previously in the cooler climates, and they are therefore more susceptible to its narcotic effect. In cooler lands the early excitement phase following the taking of alcohol is more marked than in warm climates and therefore more liable to lead to violent behaviour. There is also the point that while inhibitions are few among low-energy peoples of the tropics, they are socially necessary among the more dynamic races; and chaotic situations may arise when the brain is dulled by alcohol.

Again, it has been found that excessive smoking, like drink addiction, is less harmful to those who live in climates where the everyday stress of the vascular system is not so severe.

No one is quite sure what are the direct effects of storms on our bodies, but they do appear to be linked with sinusitis as well as with acute respiratory and rheumatic attacks. It is thought that, as atmospheric pressure falls, our tissues take up more water, swell, and then give the water off through the kidneys after the storm centre has passed. But though we may in this case be efficiently spongelike, the frequent change in tissue water content may well lower our resistance to infection. Colds and other respiratory infections are less frequent in summer as a result of reduced storminess, not, as is popularly supposed, because of the increase of temperature. In fact, prolonged warmth reduces the body's ability to fight infection, and this is certainly one factor that must be considered when anyone proposes to leave a cool climate for a permanent 'place in the sun'.

In the USA the lower half of the Ohio valley is the chief sinusitis zone, while the northern part of the state of Ohio is known to be particularly rheumaticky. Another black area for sinusitis is the upper half of the Mississippi basin from the Rocky Mountains to the Appalachians, where all respiratory troubles are common. Over the stormy middle temperate latitudes as a whole the number of rheumatic and respiratory infections is greater than all the other diseases combined, though the stormier parts of the tropics are little better. Appendicitis attacks are likewise frequent in stormy regions of both high-latitude and tropical countries.

Broad regions of the tropics that are free from storms and depressing heat are beneficial to anyone suffering from these diseases, and in the USA a favourable area in this respect is the south-west, including New Mexico, Arizona, and much of California. But sinus troubles are aggravated by bathing and by salt spray, so that localities with coastal fog and sea breezes must be avoided if possible. Victims of rheumatic infections also do well in this dry warm belt, where damage to the heart and joints may gradually repair itself.

Since the war a very prolonged and highly successful series of advertising campaigns has been put into operation by travel promoters to induce those who live in stormy winter regions to take a holiday in the sun. British and other sun-starved Europeans may take a Mediterranean cruise, but there are, of course, many other alternatives. The exercise can be very beneficial, and if one has been away for a week or a fortnight it is not likely that the system will have become so adapted to the new found warmth that to return to more rigorous conditions presents any undue risk—not, anyway, for the normally healthy. But a stay of a month or more in a warm winter climate is a different matter. By that time the body's vitality declines and resistance to infection is sharply lowered. Exactly the same problem faces the US citizen who lives in one of the middle or northern states and takes a long winter holiday in Florida or southern California.

Leprosy is a common disease in the humid tropics, and it spreads quickly. Another is tuberculosis, because people have little resistance to it. Anyone who has had this disease, no matter how inactive it may have become, should never plan a stay in any of the moist warm regions of the world. The lowered vitality of people living in hot climates is now widely appreciated, however, and doctors send patients to more invigorating regions at high altitudes. To send them straightaway into a cooler but stormy region such as Britain during the winter could have fatal effects, since this will add to the dangers of respiratory infection and more than offset any advantage.

During the First World War tuberculosis as well as pneumonia and other acute infections proved to be fatal for many native North African troops who were not climatically adapted to endure the wet and cold of northern France. On the other hand, with French, British, and American soldiers these diseases generally took slower and less fatal courses.

Links between climatic factors and the incidence and recurrence of various diseases are interesting, because, once established, they enable people susceptible to a particular complaint to plan a move to a more favourable climate if circumstances permit. Even cancer, which is still one of the major threats of this age, is more common in high latitudes than in the tropics, particularly in elderly people. Comparisons made between various age-groups have found that the cancer death rate among the aged is 50 per cent lower in warm latitudes than in the world's cold, changeable regions. Cancer in mice acts similarly, proving that the disease is to a large extent suppressed in tropical heat.

In the northern United States, and in parts of Canada and Europe, severe heatwaves can build up rapidly during the summer months, and they are a regular seasonal occurrence in places such as the Punjab, Sind and North-west Provinces of India, Iraq and the dry, hot parts of Africa and Australia. They can also occur in places of damp heat such as the Persian Gulf, the west coast of Africa, Burma, and Malaya. In all these regions heatstroke is common, and people living in the higher continental latitudes are particularly vulnerable as a result of their normally high metabolic rate; they cannot subdue their inner fires quickly enough to meet the sudden need for heat loss. One of the world's worst regions for heatstroke, both for humans and animals, is the upper half of the Mississippi River Basin, but even in Britain the occurrence of a sudden hot spell will lead to a certain number of heatstroke cases.

It is for this reason that one needs to take extra care during the first few days of a hot spell, and alcoholic drinks should be avoided or used very sparingly—the more so in elderly people and those who suffer from high blood pressure or hardening of the arteries. Even the small percentage of alcohol in beer has its dangers in such conditions, and intoxication is much quicker if the body tissues have been dried out by previous perspiration. The point to bear in mind is that prevention is always better than cure, because one attack of heatstroke or prostration leaves an individual more susceptible to similar troubles in subsequent heatwaves, and victims must be very careful to avoid future exposure.

It has been remarked that city dwellers suffer more during severe heatwaves than do residents of rural areas, and it is certainly true that heat trapped inside buildings where there is no air conditioning may cause indoor temperatures to rise higher with each succeeding day of heat, whereas in rural areas the open ground tends to lose each day's heat through the process of radiation. However, the non-adapted person may find the heat more of a problem when in the country, unless the region is well wooded, than in town, for here it is normally possible to find an area of unbroken shade on one side of a street.

The problems brought by the large-scale movement of troops during and since the last war have led to a very marked increase in the study of acclimatisation. A world-wide exercise known as IBP (International Biological Programme) has now come into operation, and one section of this is being devoted to the study of human adaptability and involves the detailed investigation of large numbers of people, measuring their food intake and energy expenditure as well as their genetic constitution, body size and shape and physiological characteristics—all in relation to their environment. If we can find out how primitive tribes living in harsh conditions have adapted and survived over the centuries, we shall then possess information of considerable practical value.

Of special concern in this respect are temperature and altitude, the adverse effects of the latter being caused by the lower barometric pressure at high levels which reduces the pressure of the atmospheric oxygen and therefore the rate at which it can be absorbed by the lungs and passed into the bloodstream. The highest permanently inhabited region of the world is found in the Andes of South America; the altitudes here are as much as 18,000 ft, and the natives themselves possess unusual constitutions. Their powerful barrel-like chests are evidence of this.

Lowlanders can adapt themselves for a stay in high-altitude conditions, but in the first place the ascent should be gradual and all physical activity limited so as to prevent exhaustion. Then, as a result of partial acclimatisation, there will be an increased haemoglobin content in the blood which will aid the body's oxygen transport system. However, this process also raises the blood's viscosity and puts an extra load on the heart. It is therefore not surprising that there are a number of illnesses that can be attributed to life at high altitudes, including acute pulmonary oedema, which has also been the cause of death in a number of mountaineers.

Obviously, the greater the change in the height of one's environment, the greater the care necessary, and permanent residence in even moderately high altitudes, such as one finds over much of the African continent south of latitude 10° S, is not advisable for migrants without returning at intervals to lower levels for a reasonably prolonged period of rest. Much of the African plateau is over 5,000 ft above sea level, so is the vast plateau of south-western USA between the Sierra Nevada and the Rocky Mountains. The latter area includes much of the states of Nevada, Utah, Colorado, New Mexico, and Arizona. The high country extends far to the south through Mexico to the Bolivian plateau of South America.

Life at high altitudes often raises other problems, not least being a hostile environment due to cold or to violent contrasts of temperature, both of which are so bad for respiratory diseases, though one has to climb very high in tropical zones to experience cold conditions all the year round. The Spaniards came across the altitude problem when, after conquering South America, they established a capital city at Potosi, in the Bolivian Andes, at a height of 13,000 ft. Fertility here proved to be very low, so the capital was eventually transferred to Lima on the Peruvian coast.

Can man become acclimatised to cold? The IBP is making a special study of this problem, and it has been proved that rats, rabbits, and guinea pigs have been able to adapt themselves to cold but, at present, man's own attempts have not been very fruitful. He has learnt how to live in the world's bitterest climates by using special clothing, which of course successfully limits real adaptation; this applies to the Eskimos as much as to Polar explorers, except that the Eskimos' hands and fingers have become climatically adapted to their environment, enabling them to carry out skilled tasks with the hands despite very low temperatures.

Two primitive tribes who, in different ways, have adapted themselves to cold, are the Australian aborigines and the Acalufe Indians of Tierra del Fuego, the group of islands forming the southernmost tip of South America and lying between the Strait of Magellan and Cape Horn. The aborigine is of special interest, for he uses no clothing to protect himself despite the fact that for many months of the year night temperatures in parts of Australia are close to freezing. Aided only by low windbreaks and a few small fires, he sleeps naked on the ground. No Europeans have been known to endure this without being acutely uncomfortable. By day, even during the cold season, the radiation received from the sun enables the aborigine to warm up quickly.

The Indians of Tierra del Fuego, who, like the aborigines, are now dwindling in number, endure one of the most miserable climates on earth. It is almost perpetually overcast, with driving winds that bring rain, sleet, and snow, yet Darwin, on his visit here, was impressed by the sight of a mother who was sitting out of doors, nursing her baby, with snow falling on her bare shoulders. Others swam in water close to freezing. One asset that these people possess is a high heat production throughout the night, which enables them to maintain an almost constant body temperature. Anyone else who wished to adapt himself to the environment would need to experience a physiological change so that he increased his internal heat production and cut down heat loss through the skin.

That man can adapt himself to live in a hot climate is well known. The British do so during the rather rare good summers that we experience in this country when the temperatures correspond roughly to those of the sub-tropics. The process is of course carried out with the aid of the sweat glands, and water that is transpired from the skin surface has a pronounced cooling effect on the body. The process has actually been studied in the laboratory, using special climatic chambers in which subjects can be exposed to precisely controlled conditions. But a first exposure is likely to give the subject some discomfort, for the sweat rate is then relatively low and results in the body temperature rising too quickly. However, the interesting point about this experiment is that complete acclimatisation to heat can be obtained by relatively brief daily exposures, with the rest of each day being spent in cooler conditions. Many different people have been tested in this way in all parts of the world, and all have shown the ability to acclimatise.

In addition to knowing how we can adapt ourselves, if necessary, to changes of climate, it is also important to discover how the human frame reacts to the short-term elemental changes that we in Britain know as weather. We share these experiences in some ways with other countries, though there are differences: for example, they are less marked in the temperate regions of south-east Australia; similar over New Zealand, though here there is not the equivalent of the cold continental influence that brings occasional severe spells to Britain; generally more extreme, and with shorter seasonal transitions, over the northern USA, northern Japan and over much of northern and central Europe; while in southern Chile the weather changes are dominated by a pattern of greater rainfall extremes.

Now when Boswell drew Samuel Johnson into a discussion concerning the best season of the year for carrying out literary work, the Doctor treated the whole notion with ridicule. But he was wrong. To be in a state of revolt against the weather is good for one's productive senses so long as the unsettled and generally stormy conditions that produce this feeling do not persist for so long—or are so violent—that we eventually feel quite exhausted—in fact, 'under the weather'.

At the beginning of the present century the American researcher Edwin Dexter explained that in terms of what he described as energy reserve. Weather, he declared, could make people inert or active; in the latter case it would enable them to draw on supplies of reserve energy that they did not realise they possessed,[2] and it has been said by more modern writers that the drop in barometric pressure and temperature before a storm can make housewives rush around and madly finish off every single job on hand, while husbands find themselves cleaning the family car despite the fact that rain is on the way and is probably mentioned in the forecasts. Perhaps this explains, at least in part, the well known grumble that it always rains after one has cleaned the car. Nevertheless, too many storms or depressing days will use up the energy reserve and lead to an increased chance of petty irritations or quarrels getting out of control.

Monotonous weather, on the other hand, that makes no demand upon the human being to do something creative or useful, has a depressing effect of a different kind, and it has been said of those who live in the steamy equatorial regions and experience conditions that hardly ever vary, that they are 'usually listless, uninventive, apathetic and improvident'.[3] Even in comparatively high latitudes a damp wind causes a feeling of lassitude, an example being the east wind that produces the Levanter cloud over Gibraltar. Even more enervating and upsetting is the hot, humid *zonda* (see p 257) of the Argentine—a northerly wind from the tropics that is well known in Buenos Aires. By the time it has reached the city 'it has become so overcharged with moisture that everything becomes instantly damp and inducing great liability to colds, sore throat and all the consequences of checked perspiration'. So remarked Sir Woodbine Parish, who went on to declare that the wind made the inhabitants temporarily deranged of their moral faculties.

Investigations into people's reactions to climate and weather reveal that these are always relative to the normal expectancy. Dexter found that in Denver, Colorado, only a few cloudy days were needed to upset the emotional balance, because here the climate for most of the year is remarkably sunny. In New York, where cloudy days are more frequent, it required a more severe change in weather to produce a similar reaction.[4]

From another comparison it was shown that the Coloradoan regards the wind as the most vulnerable point in his climate. 'These horrible winds' and the 'windiest place I ever knew' have been common local expressions, yet the windiest months in Colorado (March and April) are less windy than the calmest months in New York (July and August). The low humidities all the year in the dry south-west regions of North America, and what is known there as the 'electric state of the atmosphere' during windy periods may have a bearing on this.

In the higher latitudes of Europe and North America the winter is a regular and heavy burden to carry. The body works hard to cope with the demands made upon it by perpetual storms and changes of temperature, and all too often sunshine is at a premium. The mind has no choice but to leap from the present to focus on the approaching joys of spring, which is pictorially a season of joy and inspiration. But for all that, the demands made upon the body by the changing weather of March, April, and May are nearly as great as those of the winter. Particularly difficult to endure are the days of sudden storm and sharply falling temperature, which, because one's physical resistance is at a low ebb, bring chills or promote other infections. Then there is that other unpleasant condition known to many as 'spring fever', when one wakes up feeling that one has not slept at all and when every small task is fatiguing in the extreme.

These symptoms are caused by the body's difficulty in adjusting its metabolic rate, stage by stage, to the upward trends of temperature, and each reversal of temperature—as winter fights its rearguard actions—makes it necessary to start the process all over again. The more sudden and the fiercer the rise in temperature, the greater is the adjustment that has to be made, so that, for many people, the first really warm spells leave them exhausted rather than elated.

During the summer the most trying days are those that produce what Professor Gordon Manley describes as a 'warm grey sky',[5] and when the cloudsheet is low and has a peculiarly oppressive quality. These occur all too frequently during unsettled or changeable summers. The wind is generally light or moderate in strength, from the south-west, and the temperature 65° F or above; the higher the temperature rises the more enervating such days become. A similar but more sudden burden is placed upon the human body at the onset of thundery weather, after heat, and because most of us have difficulty in coping with a rapid humidity increase at a moderately high temperature level we suffer from a form of physical and mental strain that produces headaches and puts nerves on edge. There is also the point that very thundery weather may be accompanied by frequent though comparatively minor localised changes in barometric pressure, and they may have some effect upon the body.

Common to European countries and to others with similar climates is the rise in the suicide rate in periods of great heat, but not so much during the extreme heat of the advanced summer season as in that of the spring and early summer. The direct responsibility is not, of course, the weather's, but it is true that, under certain meteorological conditions, what is barely endurable can suddenly become unendurable. On this subject, and with reference to weather in Britain, an anonymous writer[6] declared some years ago:

> The constant gloom of bad weather ought to acquaint us so thoroughly with moods of depression that suicide would never occur to us. Look at Scotland, for instance, where suicides are rare. Why are they rare? Simply because a succession of Scottish Sundays has so accustomed the people to prolonged despondency that any sudden misfortune cannot sink their spirits any further. One has only to spend a dozen Sundays in Glasgow or Edinburgh to become inoculated against suicide. So far from November fogs driving people to jump off Waterloo Bridge, they ought to train and educate the mind to bear any calamity . . . We can educate our mental susceptibilities as we can our muscles, and the more we educate them the more they will bear.

Beneath the jocular vein of this quotation there appears to be a basic truth. The sturdy determination and evenness of temperament of the British race as a whole must be in some part due to the climate. In hot sunny countries one all too frequently finds the prevailing emotional equilibrium a delicate thing that can be touched off by a weather spark as by anything else. For those who find the British climate highly unacceptable this should come as a warning, for, often, what on first sight appears to be an ideal climate in foreign parts has its problems.

Yet there is much to be said for avoiding the stormy British winter and all the chronic ills that go with it; these may leave us largely unaffected during youth but have a nasty habit of cropping up at a later date. The comparatively warm, sunny and much drier winter climate of the Mediterranean is much better, and still more placid are the sunny Canary Islands and Madeira, situated to the south and north respectively of the 30 degree parallel of latitude. Not that temperatures here are outstandingly high during the winter, but the mean January temperature at Funchal, Madeira, of 60° F and at Las Palmas, in the Canaries, of 64° F should hardly give rise to complaints from British visitors. They are some 20° higher than the January values for most places here at home.

15

WORK FOR THE AMATEUR

METEOROLOGY makes a fascinating hobby. There are many branches to it and many things that one can do. To begin with, one can keep useful records—in fact, a miniature weather station—with a minimum of equipment, such as a barometer, a rain gauge and a set of thermometers. The latter should include an ordinary air temperature thermometer, a maximum and minimum thermometer for measuring the highest day temperature and the lowest night temperature for each 24-hour period, and, as an addition, a grass minimum thermometer for measuring minimum grass temperatures at night. Another relatively inexpensive instrument, known as a 'wet and dry bulb thermometer', and which, as the term implies, is two instruments in one, measures the air's moisture content in terms of relative humidity, and this is particularly interesting to record during the spring when showery and very dry periods tend to alternate in quick succession. It will also be of interest at any time of the year to ladies who complain that their hair is being affected by the damp weather and is becoming unruly.

At a later stage more elaborate instruments may be added, if desired, and for the fun of it anyone with a practical turn of mind can make his or her own instruments. However, equipment that is made by a recognised instrument manufacturer is an advantage because, sooner or later, every amateur finds that his records are consulted by other people or published in the local press, and accurate measurements then become essential. Appendix 12 lists the names and addresses of firms who manufacture weather instruments and who will be glad to furnish particulars and current price lists, on request. But you may be able to obtain secondhand instruments by advertising in your local journal or in the columns of *Weather*, which is published by the Royal Meteorological Society at 49 Cromwell Road, London, SW7. The same magazine also caters for announcements from people who have instruments to sell. To give just one example, in the December 1967 issue a complete 'Snowdon' pattern copper rain gauge was offered for £3, which is a little under half the price of a new gauge.

The ideal site for an amateur weather station is a small level piece of garden lawn, approximately 30 ft × 20 ft with the grass cut short at all times, and, if possible, sited not too close to buildings, trees, hedges, walls, tall fences, or anything that can interfere with sunlight and temperature and give inaccurate readings. Of course, if your garden is small, you may have to compromise somewhat, and you may not be able to find land that is absolutely level. The point here is that steeply sloping ground will modify instrument readings to some extent, particularly those provided by the rain gauge.

One other precaution is needed and that is to ensure that the direct rays of the sun do not touch your thermometers. They should actually be housed in a wooden box standing 3 ft 6 in above the ground, known as a thermometer screen, which needs to be painted white to reflect the sun's rays. A screen can be purchased from a weather instrument manufacturer or, again, you might be fortunate enough to obtain one secondhand.

For useful later additions to your weather station, which will considerably enhance the value of the combined observations, you may consider acquiring an anemometer to measure the wind speed and a wind vane. Some firms make an instrument which combines the two functions. In addition, the following instruments, though of a somewhat higher price, may eventually be acquired: a barograph, which was described on page 145, for providing a continuous record of atmospheric pressure, a thermograph for providing a continuous record of temperature, and a hygrograph for providing a continuous record of humidity. Some manufacturers produce an instrument that will combine all these functions. Then, in the luxury class, are the distant-reading instruments which operate dials that can be read inside your home, and it is possible to buy an auto-recording rain gauge, which, though not a distant-reading type of instrument, saves one from having to empty the collected water into a measuring jar each time a reading is taken.

So it is possible to spend either a comparatively few pounds or many hundreds of pounds on fitting up your personal weather station, but all records produced from it, whether simple or ambitious, will form the background for an interest that will grow as the years progress. But for anyone keeping a station the one absolute essential is the regular reading of the instruments at one or more set times each day and noting the results in a weather log. A log based on morning and evening observations is ideal, and, because breaks in observations in the log will cause annoyance when you wish to look up past weather, it is advisable to appoint a reliable person to take readings for you when you are unable to take them yourself.

Your log can be made in any reasonably large notebook; one containing foolscap pages is best, but a loose-leaf book is not advisable, for the pages will become torn in time and may fall out. The specimen sheet of the weather log, reproduced here, is one that you can follow exactly or adapt to meet your own requirements. It is important to record whether the observations are taken at Double British Summer Time, British Summer Time or Greenwich Mean Time.

The first two columns for date and time are self-explanatory, but you will note that the next column is for barometer readings. Those obtained from an inexpensive aneroid barometer will be quite satisfactory, provided the instrument has been properly corrected to allow for your own location. Any weather

SURFACE WEATHER OBSERVATIONS
(Land Station)

Entries commencing..

are for { G.M.T. / British Summer Time / Double British Summer Time

Address ....

....

....

Latitude........................ Longitude....................

Date	Time	Barometer		Thermometers		Relative Humidity	Precipitation (in inches)	Wind		Clouds		General Weather		Other Remarks
		Reading	Tendency	Air Temp. (°F.)	Wet Bulb Temp. (°F.)			Speed (m.p.h.)	Direc-tion	Amount of Sky covered	Type	Present	Last 12 hours	

Fig. 19. Specimen page of weather log for an amateur station.

instrument store or jeweller who sells barometers will be able to advise on this point.

Occasionally an amateur meteorologist will be able to earn a small additional income by submitting his observations on a regular prearranged basis to the Met Office forecasters after first having his site approved as an official reporting station (see p 48). But the site must be in a fairly remote region of the British Isles, for, elsewhere, the number of existing stations is likely to be already adequate. Apart from this, the records collected by hobbyist weathermen are received very gladly by several other organisations, and though there are seldom financial rewards for contributing the information, the exercise results in a considerable broadening of knowledge and is very worthwhile.

To give one example, the collection of climatic rainfall statistics for the British Isles—which, does not rely on the network of weather forecasting stations, but is compiled from a large number of reports from each area of the country—is done very largely by amateur observers, and there are, in fact, some 5,000 amateur rain gauge stations in Britain. Each year some observers drop out of the scheme, so more are constantly needed, particularly in areas of scanty population. The information is collected by the British Rainfall Organisation, which is now integrated with the Meteorological Office and is used for the compilation of official records.

With the aid of this basic data the Ordnance Survey, at Chessington, Surrey, has just published a new set of 10 miles to the inch rainfall maps of the British Isles for the very first time, and this is rightly described as 'invaluable to schools, business men and administrators'. All the previous rainfall maps were drawn on a much smaller scale and did not show regional variations to anything like the same degree. Of course, local rainfall as well as other climatic records are discussed in particularly great detail when it comes to building a new town, airport, reservoir, motorway, hospital, school, or any other important building.

The Meteorological Office at London Road, Bracknell, Berks, will supply any information on this matter to those who are interested, and if the observer does not already possess a rain gauge the organisation may supply one on loan.

Apart from providing the ability to contribute work of national importance—and it must be remembered that the more developed and complicated our civilisation becomes, the more sensitive it is to the effects of the weather—one of the fascinating things about becoming an amateur weather observer is that, sooner or later, one is given the opportunity to specialise.

Take thunderstorms. Observers are required to assist in a thunderstorm survey that is carried out by the Electrical Research Association at Leatherhead, Surrey. The survey provides data on which to base advice to Government departments and the local authorities on the protection of buildings and other structures against lightning. It also plays a useful part in the investigation of lightning faults on electricity supplies and helps in the development of special equipment to measure the incidence of lightning in all parts of the world.

Dr R. H. Golde, who has made a close study of lightning problems and who is in charge of the survey, draws most of his reports from some 1,600 voluntary observers in all parts of Britain. An average of 5,000 reports are sent in each year by observers on pre-paid postcards, giving the time and duration of each storm, with notes on its severity and the estimated number of lightning flashes to earth. In return, the association gives each contributing member a copy of the annual thunderstorm survey for Great Britain as soon as this is published.

There is at present a shortage of thunderstorm observers in the remoter regions of north and west Scotland (including the Scottish Isles), Cumberland, the North Riding of Yorkshire, and Lincolnshire. Extra observers are also needed in regions farther south, including the Sandringham and Dereham areas of Norfolk, west Suffolk, the Cotswold region between Stow and Oxford, the Kent coast between Margate and Whitstable, the Bishops Stortford and Buntingford areas of Hertfordshire, Pembrokeshire, the south Bristol Channel coast, and the north Devon area between Bampton and Torrington. There is also a shortage of observers in the rural areas around Birmingham and Coventry.

In a recently published report the ERA survey revealed that 1960 was a particularly noteworthy year for thunderstorms, when some 8,000 report cards were received. However, with thunderstorms, as with all other elements, surprises occur from time to time. One occurred in 1967 when Kew Observatory recorded no thunder at all for the whole of the month—the first time since August 1948. It is events like this that make meteorology such an interesting subject for amateurs as well as for the professional to study in some detail.

Occasionally the amateur weather enthusiast is able to witness something that is distinctly unusual, and for this reason he is self-trained to be constantly on the alert. In the course of time he might encounter a rare phenomenon such as a tornado, and provided that the funnel-shaped cloud actually reaches the ground, the Building Research Station at Garston, Watford, Herts, will be glad to have a report of it; this organisation is particularly interested in freak storms and in the damage that they cause. It also believes that a number of tornadoes occur each year in this country without being noticed, and more information regarding their total number and the courses that they take is urgently required. Observers should make a report of the time a tornado is seen and of the current general weather; if possible they should also mark the track of the tornado on a large-scale Ordnance Survey map. In the USA, where tornadoes are more frequent and cause more damage, the Weather Bureau sends out urgent warnings to those who are likely to be affected, but it comments that the lack of community preparedness can contribute to a high death roll.

The point one must remember about all research into

climate and weather is that, even today, there are a considerable number of unsolved problems, and while it falls to the professionals to carry out the scientific research and the design of new instruments, the field for investigation by amateur meteorologists is likely to remain very wide as far ahead as one can see.

Since the war a considerable number of new magazines have been launched to cater for those with specialised interests, and amateur meteorologists have not been neglected. Their needs are well met in this respect by the monthly journal *Weather*, even though some of the material it publishes is intended for the professional researcher. Each issue is well balanced and contains lively, often controversial articles, correspondence and queries columns, books reviews, cloud and weather photographs, news of meteorological activities from all parts of the world, and a monthly weather log giving the day-by-day pattern of the previous month's weather. From time to time the magazine also gives details of instruments that one can make and of special research requiring the co-operation of amateurs in various parts of the country. In addition, it provides details of worthwhile holiday courses in meteorology, some of which are designed with the amateur in mind. One course which has been held regularly every September for many years is the one in 'Meteorology' at Malham Tarn Field Centre near Settle, Yorkshire. The course includes a variety of practical work in and out of doors, in addition to lectures and films, and special reference is generally made to current weather. The programme normally includes excursions to nearby places of interest. Various instruments are available for outdoor use, and visiting lecturers talk on various topics during the evening.

Another very interesting course on 'Weather and Sailing', is held at Falmouth—again during September. As the title implies, it is designed for all who have an interest in both sailing and meteorology. Morning lectures throughout the week are devoted to a study of the wind, weather, and seamanship, and whenever possible lectures are linked with practical exercises in and around the Fal estuary. Afternoons are spent under sail unless the weather conditions are unsuitable.

The need for the amateur to keep up to date with his hobby by taking a regular magazine can hardly be stressed too much, and he would also enjoy another excellent publication, the monthly *Meteorological Magazine*, which is edited by a member of the Met Office staff and published by the Stationery Office. Like *Weather*, this magazine publishes a wide variety of articles, though the majority of them are not written specifically for the amateur.

Those who own their own private weather stations will find it interesting to compare their daily and monthly records with those from other parts, as well as with the information given in the Met Office forecasts (see p 82). The accuracy of the latter can therefore be fairly gauged, provided the comparison is made over a reasonably extended period and in some detail. The general public will complain bitterly if too many forecasts prove to be wrong in any given period, but the only persons really competent to criticise are those who can provide statistics to prove the points that they raise.

Evidence from local work of this kind may help the forecasters to interpret certain regional or local developments, but there is of course a limit to what they can announce within the space of a single bulletin. Showery days are particularly difficult for them, because it is impossible to say in a forecast for a comparatively large region exactly when and where the showers will occur—or if they will occur at all. Some areas will probably escape them altogether. But the local weather observer has a better chance on days like this, for his own experience and collected records may tell him that westerly winds give fewer showers than northerlies or north-westerlies; or it may be the other way about, or that no showers will occur until the wind backs to a point south of west. So much depends upon the local topography, and temperatures, sunshine, and relative wind strength may all be important considerations.

In any area a local man who has studied these matters, and

who is probably aided by a knowledge of weather lore, may therefore be able to go into competition, so to speak, with the Meteorological Office.

This has happened on a number of occasions. In 1956 the Town Council of Cleethorpes, Lincs, who were disappointed with the over-cautious Whitsun forecast of that year, appointed one of their nightwatchmen, Mr Harry Boon, to issue daily forecasts based on his lifelong observations of the flights of birds and insects, and it was agreed that from the beginning of August until the end of the summer Mr Boon's predictions should be compared with the official forecasts to see who was the more accurate.

Harry Boon won the competition. He was presented with a memento by the Mayor and given a free holiday, which he chose to spend in Cleethorpes. Mr Boon has now retired and his place has been taken by Mr Ray Cambray, the pier superintendent, whose daily forecasts during the holiday season are announced on the pier noticeboard. The Cleethorpes Director of Publicity and Entertainment states that Mr Cambray studies the winds and tides and maintains a high standard of accuracy.

Skegness is another east-coast town that has been dissatisfied with the official forecasts and has appointed its own 'met expert'. There is no doubt that this area of Britain is particularly sensitive about its weather being incorrectly reported, and spring is a critical time; it is then, as well as during late winter, that easterly winds are most frequent, and, if they blow strongly, they are likely to bring in cloud, for they will have picked up a considerable amount of moisture from the North Sea. But not all areas are affected equally or at one and the same time. As the Cleethorpes local authorities have stated:

> Our complaint with the Met Office is that our little corner of England is lumped together with the general forecast for the North. Actually this Cleethorpes area, backed by the Lincolnshire Wolds and fringed by the sea, apparently has variable weather from the forecasts—which a local expert can more accurately estimate.

THE DAILY FORECASTS

A quick check by using cloud and wind indications*

Clouds			*Wind*		*Forecast*
Type	*Movement*	*Time Seen*	*Direction*	*Strength*	
Small, rounded	Slow	Any	Any	Light	Fair, dry
Do	Fast	Early morning is most usual	West to North	Increasing	Showers soon, with squall risk
Large, rounded with dark bases	Slow	Generally late morning or afternoon	Any	Light or moderate	Showers later; thunder risk in summer
Do	Increasing, or fast	Do	Generally from West, South-West or South	Moderate or strong	Imminent heavy showers; Thunder risk in summer
Low sheet or rolls, not dark	Slow	Any	Do	Light	Dry for next 6–12 hours
Low sheet, becoming darker	Increasing	Any	Do	Moderate	Rain and strong wind imminent
High sheet	Covering only part of sky	Any	Any	Light or moderate	Remaining fair
High sheet (milky look)	Covering up whole of sky	Any	Generally from West, South-West or South	Increasing	Rain within 6–8 hours
Large, any type, during rainy spell	Any, but thinning out	Morning	Any	Generally decreasing	Becoming more settled and dry
'Mackerel' sky	Slow	Generally evening	Any	Light to moderate	Fair for next 6–12 hours

Do	Thickening or fast moving	Do	Generally from West or South-West	Increasing	Rain within next 12 hours
Parallel bars of cloud	Do	Any	Do	Do	Rain within 6–12 hours
More than one type at different levels	Slow	Any	Do	Light	Rain by following day
Do	Increasing or fast moving	Any	Any	Increasing	Heavy rain within 4–6 hours
Heavy bank tinted deep red	Any	Any (generally after dawn)	Generally from West or South-West	Any, but probably increasing	Do
Broken layer, rose tinted	Slow	Evening	Any	Light	Fair, dry
Land mist (over high or low ground)	Increasing	Evening	Generally calm	—	Do
Do	Do	Daytime	Generally South-West	Increasing	Rain and wind within 5–6 hours
Sea mist	Moving inland	Evening	Any	Light or moderate	Dry, fine next day
Do	Do	Daytime	Any	Do	Dry and fine inland
Do	Do	Any	Generally South-West or South	Strong	Rain within 2–3 hours

* This table has been designed to cover some of the more frequent cloud and wind sequences experienced in Britain

One cannot assume, from this, that other seaside areas do not face similar problems. In June 1950 a very bitter complaint, which was reported in the Press, came from the Bristol Channel region. Mr George Brenner, owner of Weston-super-Mare pier, had threatened to apply for an injunction to prevent the BBC from broadcasting weather forecasts. He claimed that after the BBC had forecast 'windy, dull and showery' weather on Whit Sunday it turned out to be a day of blue skies and 12-hour sunshine. Inaccurate prediction, he continued, 'costs seaside shopkeepers thousands of pounds'. But Mr Brenner, who was, of course, referring to the Met Office forecasts that are broadcast by the BBC, was disappointed to be advised that neither the BBC nor the Press have a legal duty to take care that their statements about the weather or anything else are correct.

As an example of errors that can occur in inland regions, a *Financial Times* correspondent reported, in September 1956:

> It is possible that the contours in my part of England, lying to the leeward of the Welsh mountains, with the Wrekin to the east and the Church Stretton range to the south, make weather forecasting a less exact science here . . .

From the point of view of the amateur researcher, reports of this nature, which are bound to occur from time to time and are not always made resentfully, do furnish very good scope for a close look at local conditions in relation to the official forecasts over a season or more. If repeated errors are discovered, no doubt the appropriate regional Met Office would be glad to know about them.

Amateur meteorologists who are also yachtsmen, or who are engaged in other outdoor sports or activities that are highly sensitive to weather, will appreciate the importance of listening regularly to the combined shipping forecasts and coastal waters bulletins that are broadcast regularly on 1,500 metres. However, a problem arises. The information is read too quickly to be taken down in longhand, and even if one uses a form of shorthand, and the forecast or its main implications are fully understood, the information thus acquired does not always present an easily recognisable 'at-a-glance' weather map.

An ingenious way of overcoming this problem has been devised by a former Met Office official now working in Australia, Mr C. E. Wallington. In his booklet 'Your Own Weather Map' he describes what, after a week or two of practice, is a relatively simple technique for constructing weather maps from the information given in the shipping forecasts. It can be obtained from the Royal Meteorological Society, 49 Cromwell Road, London, SW7, price 6s, post free. A pad of outline 'Metmaps' that have been specially designed for plotting the shipping forecasts can also be supplied by the publishers, and these are available in blocks of twenty-four maps per pad, price 7s 6d, post free.

The system was introduced in 1964 and since then its use has steadily increased. Mr Wallington makes the point that most weather maps in the Press have to be prepared too early to be of real use to the amateur forecaster and that the television presentations, though admirable, appear too late in the day for the serious amateur whose sports, hobbies, or studies depend on the day's weather. He also quotes the case of the amateur observer or cloud photographer who, while he may not actually want to make a forecast every day, finds that 'his interest is enhanced if he knows how his local view fits into the framework of the broad-scale situation, and an up-to-date pressure map can stimulate school interest in using the atmosphere as an outdoor extension to the physics laboratory'.

In the last few years much useful research has been carried out by various schools which run special activities and projects connected with weather. Some of their activities are published in *Weather* from time to time, and at one period this journal did in fact produce a popular detachable 'schools' supplement.

One particularly interesting piece of research carried out by the Archers Court Secondary School, Dover, illustrates what can be done by a group of amateurs working together. On 14 October 1964, during a generally dry month in most parts of Britain, a torrential rainstorm affected the Dover and Dymchurch regions, and its impact was such that the school's geography department began a special investigation; one of the objects was to discover if the causes were related in any way to the area's local topography.

The Dover Corporation was consulted, also many of the local schools, the Mid-Thames Water Company, the borough engineers of the coastal towns between Eastbourne and Margate, and, finally, the Meteorological Office. The school obtained 140 observations and from these constructed a series of charts which were later analysed by experts. The conclusion was that the storm was in the nature of a chance event and was not accentuated to any marked degree by the local conditions.

For a number of people who like to develop their own weather interests in a quiet way and who do not perhaps have the time or the opportunity to take regular instrumental readings, the keeping of a weather diary for the area in which they live provides useful information in the course of time. Long notes are not essential, and each daily entry need take only a few minutes. A retired bank manager living in the Bideford area of north Devon has been doing this since 1961, using a 5-year diary for his combined personal and weather notes. Even before his first diary was completed, he found that he was constantly referring to back years for information about the state of the various seasons. At the end of 1965 it appeared that the number of fine sunny days had varied little from year to year, the figure for the area being 166, 170, 151, 162, and 152 days respectively. General weather was recorded each day as fine and sunny; fine but dull (that is, not wet); or wet (that is with rain or other precipitation). Because the weather often changed in the course of a day the diary dealt in half days where necessary, and the halves were added together at the end of each year.

Obviously a 5-year period is too short, climatically speaking, to come to any broad general conclusions, and a 10–15 year period would be more revealing. However, it was interesting to note that during the period 1961–5 in the area just west of Bideford, unsettled weather tended to predominate until about 11 June if no good spell of weather occurred during the 5-day period before or after 25 March. It was also noted that the conditions during the last week of June had a habit of influencing the weather to a considerable extent over the following three months. The Buchan periods were not particularly prominent in the area during the years in question, and the cold period 6–13 November gave a white frost in only one year in five.

This type of information shows what can be done with the minimum of effort and no outlay apart from the cost of a diary. How often one wishes to improve on one's weather memory and look back to check on what the weather did some years ago. It is always fascinating and sometimes very instructive to compare past weather with present, particularly for keen gardeners, people wanting to pick the best period for painting the exterior of a house, or others wishing to check the weather with fuel consumption.

Some amateur meteorologists specialise exclusively in rainfall recording, and, in addition to contributing information to the British Rainfall Organisation, form themselves into local groups in order to study the rainfall peculiarities of their respective areas in as much detail as possible. In a number of counties special rainfall organisations (see Appendix 11) have been formed and issue monthly reports to their members summarising the position for the previous calendar month.

Those who own rain gauges, or would like to borrow one, might find it useful to offer their services to the nearest area river board, as these authorities often need extra observing stations. The Devon River Authority, for instance, drew up a plan in 1966 showing its area of 2,415 square miles and the approximate positions where new rain gauge readings were required. Among the volunteers recruited were farmers, postmasters, owners of petrol filling stations, parsons, retired professional people, and gardeners of country estates. Very few withdrew later from lack of interest.

Those who do keep daily rainfall records may well be interested in the very convenient at-a-glance method of displaying their statistics, which is reproduced here by kind permission of its originator, Squadron Leader T. B. Norgate of Taverham, Norwich:

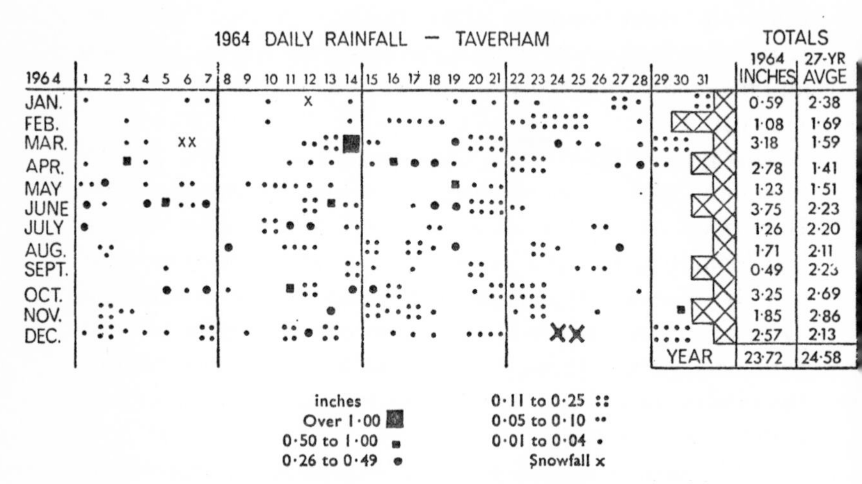

Fig. 20. A simple way of recording rainfall amounts.

Another useful line of enquiry for the amateur meteorologist is to study the effect of weather on bird migration and how efficient various breeds are at foretelling what is to come. There is room for a considerable amount of extra research on this topic. Joyce Whillis of Co Durham mentioned[1] that after several seasons of observations she was convinced that a number of species could sense coming conditions from an hour to a week in advance. Robins, during bad seasons, held their territory from November until the end of February or the beginning of March but disappeared from time to time during milder periods. Nearly always the sudden return of a robin preceded bad weather by about four days. She also noted that 'without exception the southward or south-westward movement of ducks, swans and geese has preceded a severe spell. These birds usually provide the first indication of a coming change. The geese may pass over eight days ahead of the cold weather, as they did last November'.

Another interesting observation from this writer showed that the birds were not foolish enough to be caught out by a temporary mild interlude. Between 4 to 6 February 1948 'every type of bird usually seen during a bad spell arrived in the garden'. Snow showers occurred on the night of the 6 February, and then it was milder on the following day, but the birds did not leave their sheltered territory until 9 February, after a great hurricane had swept northern England.

Press cuttings about weather are worth collecting. At the beginning of each month *The Daily Telegraph* and *The Times* are among the newspapers that provide short commentaries in their inside news columns on the weather conditions of the previous month. These can be cut out and kept in a general record book, and they can be augmented by any daily cuttings regarding unusual occurrences or new records.

How long the exceptional will remain exceptional is a question one must constantly bear in mind, for a local temperature value not exceeded for, say, 20 years may be broken again during the following year. It must also be remembered that individual dramatic events as reported in the Press may not be typical of a wide regional area, let alone for the whole of Britain.

An unprecedented event in Britain, witnessed from nearly every part of the country as well as in Scandinavia, Germany, and Switzerland, and, not surprisingly, not since repeated, was the appearance of a blue sun, and, shortly afterwards, a blue moon, in September 1950. An RAF aircraft reported finding a layer of dust and sand at 43,000 ft over the east of Scotland, which could have been the cause. It was thought to have spread from forest fires that had broken out in several parts of Canada and had been carried eastwards by the prevailing winds at high levels. As *The Daily Telegraph* pointed out, the basis for the adage 'once in a blue moon' had at last become obvious.

Apart from giving the amateur meteorologist a growing source of background information, press cuttings dealing with various aspects of the subject can also provide him with amusement. Examples are more frequent than one might suppose, but very often the humorous element is provided by creating a sequence of cuttings rather than by looking at a single one. Easter 1951 provides an excellent example, as the following extracts clearly demonstrate. It was an early festival that year, with Easter Sunday falling on 25 March:

> With luck m'lady should be able to don a new bonnet. Rain or drizzle is forecast for tonight and early tomorrow, but long sunny intervals and generally fine weather is the 'further outlook'.
> (*Evening News*, 21 March)
>
> Two Easter Question Marks: Weather and the French Strike.
> (Headline from column in the *Evening News*, 22 March)
>
> Deep depression over Baltic moving slowly East. Rain spoils holiday start. (*Daily Telegraph*, 24 March)
>
> Brilliant sunshine, blue skies . . . yesterday morning nearly everywhere . . . but millions missed it. By the time they had really stirred themselves in many parts of the country the skies had begun to cloud. (*Daily Graphic*, 26 March)
>
> Inch of snow in the north (*Daily Telegraph*, 26 March)
>
> Sun due now, say the Weather Men.
> (*Evening News*, Tuesday 27 March)

And so to work.

CHAPTER REFERENCES

Chapter 1

1 p 12 O. G. Sutton. *Understanding Weather.*

Chapter 3

1 p 41 More precise specifications are as follows:
South Cone: Gale from south-east, veering to south-west, west, or north-west; or from south-west, veering to west, or north-west; or from west, veering to north-west; or from east, veering to south, or south-west.
North Cone: Gale from south-east, east, or north-east, backing to north; or from north-west, veering to north, north-east, or east; or from north, veering to north-east, or east; or from north-west, veering to east.
(The wind is said to veer when its change of direction is clockwise, and to back when this is anticlockwise.)

Chapter 8

1 p 119 R. Inwards. *Weather Lore.*

Chapter 10

1 p 148 *Weather* magazine. July 1949, July 1950, and December 1955.
2 p 148 Mr H. H. Lamb of the Meteorological Office, writing in *Weather*, January 1965, reports a decrease in westerly-type weather since the early part of the present century; there has been a compensating increase in northerly, north-westerly, and southerly types. These changes, though striking, are not indicative of any marked changes in the British seasonal structure.

Chapter 11

1 p 162 C. E. P. Brooks. *British Floods and Droughts.*
2 p 163 *Weather* magazine. November 1965.
3 p 165 W. A. L. Marshall. *A Century of London Weather.*

4 p 166 *Weather* magazine. April 1967.
5 p 167 G. Kimble. *The Weather.*
6 p 168 O. G. Sutton. *Understanding Weather.*

Chapter 13

1 p 190 'Fog' Abatement Project: Progress Report, September 1966–October 1967.' New Jersey Department of Transportation.
2 p 192 *The Countryman.* Autumn 1950.
3 p 196 Ordinary rocketry aims to break up cloud hail pockets purely by high explosive. Silver iodide seeding aims to materially increase the number of hailstone embryos so that the subsequent growth of each one is diminished.
4 p 199 'Weather and Climate Modification: Problems and Prospects.' Report by the Panel on Weather and Climate Modification to the National Academy of Sciences, Washington DC, 1966.
5 p 199 An article showing how growers can exploit local natural advantages of climate, aside from efforts to control the environment, was presented in a symposium on agricultural meteorology at the University College of Wales on 9 March 1966, and was reprinted in *Weather*, March 1967.
6 p 200 See 'Introducing Outdoor Irrigation'. Ministry of Agriculture Advisory Leaflet No 487, and 'Irrigation', Bulletin No 138, third edition, 1962.

Chapter 14

1 p 207 Professor Clarence A. Mills, *Climate Makes the Man.*
2 p 217; 3 p 218, 4 p 218 E. G. Dexter *Weather Influences.*
5 p 219 Professor Gordon Manley, *Climate and the British Scene.*
6 p 220 *Once a Week*, Vol 19.

Chapter 15

1 p 238 *The Countryman.* Autumn 1948.

BIBLIOGRAPHY

Atkinson, B. W. *The Weather Business*. Aldus, 1968

Barrett, E. C. *Viewing Weather from Space*. Longmans, 1967

Barry, R. G. & Chorley, R. J. *Atmosphere, Weather and Climate*. Methuen, 1968

Battan, L. J. *Cloud Physics and Cloud Seeding*. Heinemann, 1965

Bilham, E. G. *The Climate of the British Isles*. Macmillan, 1938

Bone, S. *British Weather*. Collins, 1946

Brazell, J. H. *London Weather*. HMSO, 1968

Brooks, C. E. P. *British Floods and Droughts*. Benn, 1928; *Climate in Everyday Life*. Benn, 1950; *Climate through the Ages*. Benn, 1949; *The English Climate*. English Universities Press, 1954

Bush, R. *Frost and the Fruitgrower*. Cassell, 1945

Dexter, E. G. *Weather Influences*. Macmillan, 1904

Edholm, O. G. 'Problems of Acclimatisation in Man', a paper read before the Royal Meteorological Society on 5 January 1966 as the Margary Lecture for 1966

Fox, R. F. & Lloyd, W. E. B. *British Health Resorts* (Eighth edition). British Health Resorts Association, 1950

Hawke, E. L. *Buchan's Days*. Lovat Dickson, 1937

Inwards, R. *Weather Lore*. Rider & Co, for the Royal Meteorological Society, 1950

Kendrew, W. G. *Climatology*. Clarendon Press, Oxford, 1949; *The Climates of the Continents*. Clarendon Press, Oxford, 1961

Kimble, G. *The Weather*. Penguin Books, 1951

Lamb, H. H. *The Changing Climate* (Selected Papers). Methuen, 1966

Lane, F. W. *The Elements Rage*. David & Charles, Newton Abbot, 1966

Laughton, C. & Heddon, V. *Great Storms*. Allen & Co, 1927

Lester, R. M. *The Observer's Book of Weather*. Warne, 1955

Ludlam, F. H. & Scorer, R. S. *Cloud Study*. Royal Meteorological Society, 1960

Manley, G. *Climate and the British Scene*. Collins, 1952

Marshall, W. A. L. *A Century of London Weather*. HMSO, 1952

Mason, B. J. *Clouds, Rain and Rainmaking*. University Press, Cambridge, 1962

Meteorological Office. *Climatological Atlas of the British Isles*. HMSO, 1952; *Cloud Types for Observers*. HMSO, 1962; *Course in Elementary Meteorology* (Part I). HMSO, 1962; *Meteorological Glossary*. HMSO, 1963; *Meteorological Observer's Handbook*. HMSO, 1966; *Meteorology for Mariners*. HMSO, 1967; *Pictorial Guide for the Maintenance of Meteorological Instruments*. HMSO, 1963; 'Sectional List No 37'—a list of meteorological and allied publications available from HMSO. Issued free. HMSO, 1967; *Your Weather Service*, HMSO, 1959; *Weather in Home Fleet Waters* (3 Vols). HMSO, 1964; *Weather in the Mediterranean* (2 Vols). HMSO, 1962

Miller, A. Austin. *Climatology*. Methuen, 1950

Mills, C. A. *Climate makes the Man*. Gollancz, 1946

Ministry of Agriculture. *The Farmer's Weather* (Bulletin No 165). HMSO, 1964

Ministry of Housing & Local Government. *Average Annual Rainfall 1916–50*. Ordnance Survey, Chessington, 1967

Newton, H. W. *The Face of the Sun*. Penguin, 1958

Panel on Weather and Climate Modification, 'Report on the Problems and Prospects of Weather and Climate Modification'. National Academy of Sciences, Washington DC, USA, 1966

Peter, N. L. *Weatherwise*. Pergamon, 1964

Pilkington, R. *The Ways of the Air*. Routledge, 1961

Rantzen, M. J. *Little Ship Meteorology*. Jenkins, 1961

Scorer, R. *Air Pollution*. Pergamon, 1968; *Cloud Studies in Colour*. Pergamon, 1967

Smith, L. P. *Seasonable Weather*. Allen & Unwin, 1968; *Weather Studies*. Pergamon, 1966

Sutcliffe, R. C. *Weather and Climate*. Weidenfeld & Nicolson, 1966

Sutton, O. G. *The Challenge of the Atmosphere*. Hutchinson, 1962; *Understanding Weather*. Penguin, 1967

Tinn, A. B. *This Weather of Ours*. Allen & Unwin, 1946

Wallington, C. E. 'Your Own Weather Map'. Royal Meteorological Society, 1967

Watts, A. *Instant Weather Forecasting*. Adlard Coles, 1968; *Wind and Sailing Boats*. Adlard Coles, 1965

Watts, W. H. *Weather for Yachtsmen*. Bosun Books, 1965

Whistlecraft, O. *The Climate of England*. Longmans, 1840

White, G. W. *Weather Maps and Elementary Forecasting*, Kandy Publications, 1967

World Meteorological Organisation. 'Weather and Food: Freedom from Hunger Campaign, Basic Study No 1'. WMO, Geneva, 1962

CLOUD PHOTOGRAPHS AND WALL CHARTS

Twenty-four postcard-size reproductions from the Cave and Clarke collection of Cloud Photographs. Royal Meteorological Society, 1967

The Weather: A series of Wall Charts. Educational Productions Ltd, 1964

FILMSTRIPS

'Air Pollution'. In two parts, 32 frames each, 35 x 24 mm, with notes by R. S. Scorer. Diana Wyllie Ltd, 1963

'A Screen Colour Guide to Clouds'. 48 colour frames and four diagrams, 35 x 24 mm or 24 x 18 mm. Diana Wyllie Ltd, 1963

'Cloud Forms'. In two parts. 45 colour frames and three diagrams in all, 35 x 24 mm or 24 x 18 mm, with notes by R. S. Scorer and A. B. Fraser. Diana Wyllie Ltd, 1968

'Cloud Recognition'. 12 cloud frames and three diagrams—a shorter filmstrip for newcomers to meteorology, with notes by R. S. Scorer and J. B. Andrews. Diana Wyllie Ltd, 1968

'Meteorological Instruments'. Colour frames and diagrams, 35 x 24 mm or 24 x 18 mm, covering a range of instruments, with notes on their design, use and siting, sponsored by C. F. Casella and Co Ltd. Diana Wyllie Ltd, 1968

'Optical Phenomena and Storms'. Two sets of 10 2 x 2 in frames, with notes by J. Paton and R. S. Scorer. Diana Wyllie Ltd, 1963

'Running a School Weather Station'. In two parts, 33 and 28 frames respectively, 35 x 24 mm or 24 x 18 mm, with teaching notes and a foreword by the Director-General of the Meteorological Office. Diana Wyllie Ltd, 1963

'Stable Weather'. 15 frames 35 x 24 mm, with notes by R. S. Scorer. Diana Wyllie Ltd, 1963

'Unstable Weather: the Growth of Cumulonimbus'. 15 frames 35 x 24 mm. Diana Wyllie Ltd, 1963

APPENDIX 1

AVERAGE TEMPERATURES, FROSTS, SUNSHINE, RAINFALL, SNOWFALL AND HUMIDITY FOR THE BRITISH ISLES

The following tables are divided into nine meteorological areas according to the numbers on the map. It must be emphasised that there is no complete uniformity within each area, and that each table gives the average.

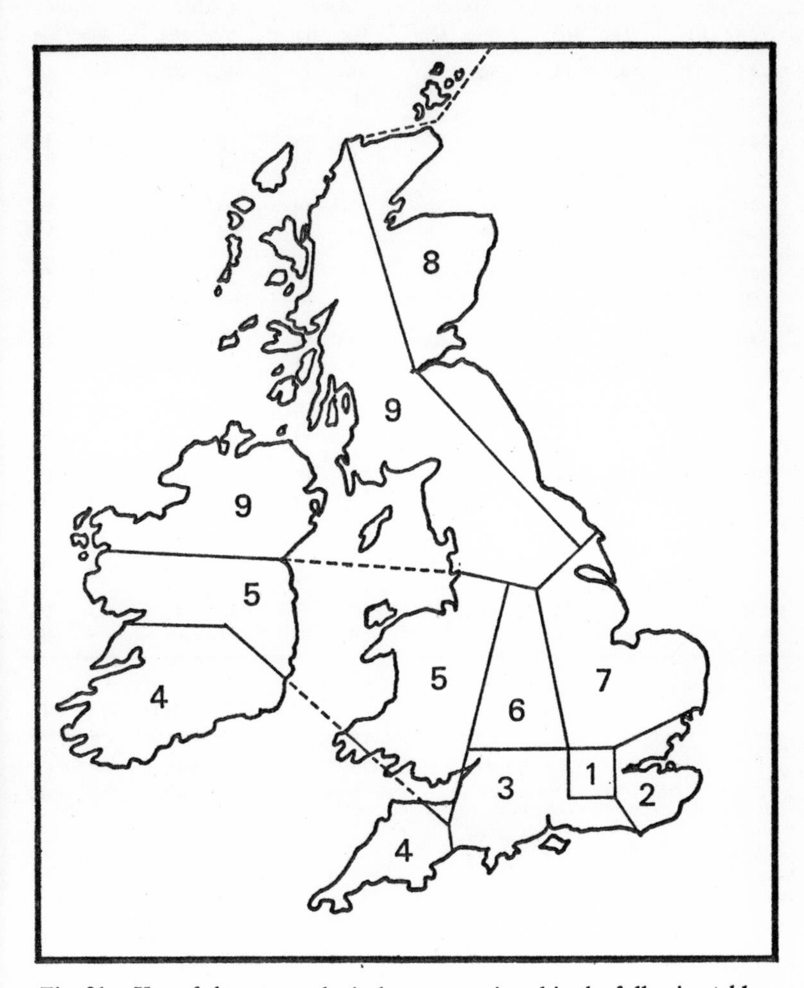

Fig. 21. Key of the meteorological areas mentioned in the following tables.

AVERAGE

Met Area No	*Jan*		*Feb*		*March*		*April*		*May*		*June*	
	Max	*Min*	*Max*	*Min*	*Max*	*Min*	*Max*	*Min*	*Max*	*Min*	*Max*	*Min*
1	45	35	46	35	50	36	55–6	39	64–5	45–6	68–9	50–51
2	44–5	34–5	45	34–5	48–50	35–7	52–5	38–40	58–64	44–7	63–8	49–52
3	45–6	35–9	46	35–8	49–50	36–9	53–5	38–42	60–64	44–7	64–8	49–52
4	46–50	35–44	47–9	35–43	50–51	36–43	53–5	39–44	58–63	45–8	62–7	49–52
5	46–7	35–9	46	35–8	48–9	36–9	52–5	38–42	57–62	44–7	61–7	48–51
6	45–6	35–6	45–6	35–6	49–50	36–7	54–5	38–40	62–3	44–6	66–7	48–50
7	45	34–5	45–6	34–5	49–50	35–6	54–5	38–9	62–4	44–5	66–7	49–50
8	44	34–5	44	33–5	47	34–6	51–2	36–8	55–9	42–3	61–5	46–8
9	44–6	34–6	45	34–5	48	35–7	52–4	37–40	58–61	42–5	63–5	46–50

AVERAGE NUMBER OF

Met Area No	*September*		*October*		*November*		*December*		*January*	
	Day	*Night*	*Day*	*Night*	*Day*	*Night*	*Day*	*Night*	*Day*	*Night*
1	—	2	—	4	—	4	—	6	1	7
2	—	1	—	3	—	3	—	6	1	7
3	—	—	—	2	—	3	—	3	1	6
4	—	—	—	2	—	1	—	2	1	5
5	—	—	—	2	—	3	—	4	1	7
6	—	2	—	5	—	6	—	9	1	10
7	—	2	—	5	—	7	—	9	2	10
8	—	2	—	4	—	5	—	9	1	8
9	—	—	—	2	—	3	—	3	2	5

TEMPERATURES IN °F

July		*August*		*Sept*		*Oct*		*Nov*		*Dec*	
Max	*Min*	*Max*	*Min*	*Max*	*Min*	*Max*	*Min*	*Max*	*Min*	*Max*	*Min*
72	54	71–2	53–4	67	49–50	58	44	50	38	46	36
67–72	53–5	67–71	52–5	63–6	48–52	56–8	43–6	49	36–40	45	35–7
68–72	53–6	68–71	52–6	65–6	48–53	58	44–8	50–51	37–42	46–8	36–40
65–70	53–6	66–9	53–6	64–5	48–54	58	44–51	50–53	38–46	47–50	36–45
64–70	52–4	63–8	52–5	61–4	48–52	56–7	43–8	50	37–43	47	36–41
69–71	52–4	68–9	52–3	64–5	48–	57	43–5	49–50	37–9	46–7	36–37
70–72	52–4	68–70	52–3	65–6	48–9	57–8	43–4	49	36–8	45–6	35–6
64–7	50–51	63–6	50–52	61	46–8	54–5	41–3	47	36–8	44–5	34–6
66–8	51–3	65–7	50–54	62–3	47–50	56	42–5	48–49	36–9	45–6	35–7

DAY AND NIGHT FROSTS

February		*March*		*April*		*May*		*Year*	
Day	*Night*	*Day*	*Night*	*Day*	*Night*	*Day*	*Night*	*Day*	*Night*
1	9	1	6	—	4	—	3	3	45
2	9	1	7	—	4	—	2	4	42
1	6	—	5	—	3	—	2	2	28
1	4	—	3	—	1	—	—	2	18
1	6	—	5	—	3	—	—	2	30
2	10	1	6	—	5	—	3	4	56
2	9	2	7	—	6	—	4	6	59
3	7	2	7	—	5	—	5	6	52
2	5	—	4	—	2	—	2	4	26

AVERAGE HOURS

Met Area No	*January* Total	*January* Daily Total	*February* Total	*February* Daily Total	*March* Total	*March* Daily Total	*April* Total	*April* Daily Total	*May* Total	*May* Daily Total	*June* Total	*June* Daily Total
1	34	1·11	51	1·82	104	3·37	133	4·42	183	5·92	198	6·59
2	54	1·75	73	2·57	130	4·20	162	5·39	210	6·76	226	7·51
3	62	2·01	80	2·83	137	4·43	175	5·85	214	6·89	230	7·66
4	57	1·83	78	2·75	138	4·47	177	5·89	206	6·64	222	7·41
5	55	1·76	71	2·51	124	4·00	154	5·15	195	6·27	193	6·44
6	42	1·35	60	2·12	104	3·34	135	4·52	177	5·70	188	6·26
7	49	1·59	67	2·36	119	3·84	146	4·86	190	6·11	204	6·80
8	53	1·72	77	2·72	116	3·73	144	4·79	178	5·75	189	6·29
9	40	1·28	63	2·22	106	3·44	136	4·55	184	5·94	189	6·29

AVERAGE INCHES OF RAINFALL AND

Met Area No	*January* Inches	*January* Rain Days	*February* Inches	*February* Rain Days	*March* Inches	*March* Rain Days	*April* Inches	*April* Rain Days	*May* Inches	*May* Rain Days	*June* Inches	*June* Rain Days
1	2·00	15	1·85	13	1·95	15	1·95	12	1·90	12	2·15	10
2	1·90	17	1·80	15	1·85	16	1·80	14	2·05	12	2·00	11
3	2·80	18	2·50	13	2·45	17	2·25	14	2·30	12	2·45	12
4	4·85	22	4·20	18	3·50	14	2·60	12	2·40	11	2·55	13
5	4·95	22	3·85	18	3·90	17	3·10	14	3·05	14	3·05	13
6	3·80	17	3·50	14	3·55	14	2·80	13	2·85	12	1·90	9
7	1·85	18	1·80	14	1·85	16	1·80	12	2·00	12	2·15	10
8	2·40	17	2·35	15	2·50	15	2·35	13	2·40	13	2·25	13
9	5·80	22	4·40	18	4·10	19	3·80	15	3·60	16	3·55	13

* Excluding days when precipitation amounts are too small to be measured.

OF SUNSHINE

July Total	July Daily Total	August Total	August Daily Total	September Total	September Daily Total	October Total	October Daily Total	November Total	November Daily Total	December Total	December Daily Total	Year Total
192	6·20	181	5·82	136	4·53	95	3·06	42	1·39	29	0·94	1378
220	7·11	201	6·48	152	5·08	112	3·63	64	2·13	51	1·65	1655
211	6·80	203	6·55	152	5·08	115	3·72	72	2·39	60	1·92	1711
185	5·98	186	5·99	145	4·83	113	3·66	70	2·34	52	1·67	1629
153	4·95	159	5·14	129	4·31	98	3·16	61	2·03	47	1·50	1439
172	5·56	164	5·31	122	4·07	93	2·99	49	1·65	37	1·19	1343
190	6·14	180	5·80	139	4·64	105	3·39	58	1·93	41	1·33	1488
156	5·04	151	4·88	130	4·32	102	3·27	64	2·15	45	1·46	1405
156	5·04	147	4·74	113	3·77	85	2·75	52	1·73	33	1·06	1304

AVERAGE NUMBER OF RAIN DAYS*

July Inches	July Rain Days	August Inches	August Rain Days	September Inches	September Rain Days	October Inches	October Rain Days	November Inches	November Rain Days	December Inches	December Rain Days	Year Inches	Year Rain Days
2·30	13	2·20	13	2·05	11	2·20	14	2·20	16	2·25	17	25·00	161
2·20	12	2·30	14	1·80	12	2·30	16	2·20	18	2·25	20	24·45	177
2·50	16	2·65	15	1·85	12	3·55	16	3·15	16	3·85	18	32·30	179
2·80	13	3·35	16	3·35	14	5·25	18	5·70	19	6·10	22	46·65	192
3·60	16	4·90	18	3·80	15	5·90	20	5·75	19	6·10	21	51·95	207
2·70	12	3·00	13	2·40	12	3·80	16	3·75	15	3·90	18	38·95	165
2·05	12	2·15	12	1·90	12	2·15	15	2·05	15	2·10	20	23·85	168
3·05	14	3·65	16	2·40	14	3·50	16	2·70	18	2·90	17	32·45	181
3·65	17	4·25	19	4·25	19	5·40	20	5·65	21	5·75	22	54·20	221

AVERAGE FREQUENCY OF SNOWFALL OVER LOW GROUND

Met Area No	*November*		*December*		*January*		*February*		*March*		*April*		*Year*	
	Days with Snow Fall	*Days with Snow Lying*	*Days with Snow Fall*	*Days with Snow Lying*	*Days with Snow Fall*	*Days with Snow Lying*	*Days with Snow Fall*	*Days with Snow Lying*	*Days with Snow Fall*	*Days with Snow Lying*	*Days with Snow Fall*	*Days with Snow Lying*	*Days with Snow Fall*	*Days with Snow Lying*
1	1	0	1	1	4	3	4	3	3	2	2	1	15	10
2	1	0	2	1	3	2	3	3	2	1	2	1	13	8
3	0	0	1	1	2	1	2	2	2	1	1	1	8	6
4	0	0	1	1	2	2	2	1	1	0	0	0	6	4
5	1	0	2	1	2	2	2	2	1	1	0	0	8	6
6	1	1	3	3	4	4	4	3	3	2	2	1	17	14
7	1	1	2	2	3	3	4	4	4	2	3	0	17	12
8	2	1	4	4	5	5	5	4	6	3	2	2	24	19
9	1	1	3	3	5	3	4	4	4	3	3	1	20	15

AVERAGE MONTHLY RELATIVE HUMIDITY

Met Area No	*Jan* %	*Feb* %	*March* %	*April* %	*May* %	*June* %	*July* %	*August* %	*Sept* %	*Oct.* %	*Nov* %	*Dec* %
1	85	80	68	65	60	60	60	68	65	75	80	82
2	85	80	75	72	70	70	70	75	70	75	80	85
3	85	80	70	70	68	68	70	70	68	75	80	85
4	85	80	75	75	75	70	75	75	75	80	80	85
5	85	80	80	75	72	75	75	75	75	80	85	85
6	78	75	70	60	60	60	60	60	65	70	80	80
7	85	85	70	65	65	65	63	66	68	70	82	85
8	80	78	75	70	72	72	72	72	74	74	78	80
9	80	80	75	70	65	65	70	72	72	75	80	82

APPENDIX 2

NOTABLE REGIONAL AND LOCAL WINDS THROUGHOUT THE WORLD

When air is forced to rise in the atmosphere—for example, by blowing against the side of a range of hills or mountains—it expands. This is because the pressure of the atmosphere decreases with height. But as air expands it becomes colder, and this process may lead to the formation of cloud; sometimes the cloudbanks will be very extensive and will produce considerable amounts of rainfall.

The opposite occurs when air that has risen in this way is forced to descend to lower levels in the lee of a mountain range. It then becomes compressed, much drier, and much warmer. The process is responsible for creating what are known in central Europe and elsewhere as *föhn* winds, which occur chiefly over northern Alpine valleys. The force that starts them moving is a depression travelling parallel to—but north of—the mountain range and so drawing in air from the south of it. This airflow drops its rainfall as it chills during its ascent of the southern Alpine slopes but becomes relatively very dry and hot as it descends the northern slopes. A descent of 5,000 ft produces a temperature rise of about 25° F.

The *föhn* blows most strongly when it descends those valleys whose courses coincide with the general flow of the wind: among the principal ones are the north and south valleys such as the Rhine from Chur to Lake Constance, the Aar to Lake Brienz, and the Rhône from Martigny to Lake Geneva. The arrival of the *föhn* is seen by a crest of cloud along the mountains to the south, which appear very close and tinged with blue. After a few puffs of cool wind, and then a short period of ominous calm, comes the *föhn's* hot blast. At its onset the thermometer rises some 10–15° F and snow vanishes in the hot dry air. Everything becomes tinder dry and highly combustible; all fires have to be extinguished, and a *föhn* watch, as a precaution against fire, is a statutory obligation on every householder.

In the autumn the *föhn* helps to ripen the harvest, and in some places the grape harvest is dependent on it. Another advantage of the *föhn* is that in spring it very quickly denudes fields and pastures of snow and hastens the start of the agricultural year.

In northern Switzerland there is an average of 41 days of *föhn* winds each year—the seasonal distribution being: spring—17, summer—5, autumn—10, and winter—9. Farther south, and in the Italian Alpine valleys, the *föhn* is less common, and the temperature rises are proportionately smaller.

The *föhn* is likely to occur wherever systems of low barometric pressure create winds that pass over mountainous regions: for example, it occurs on the coast of Greenland where it has a marked influence on the winter climate. Other examples are the *samun* of Persia, which descends from the mountains of Kurdistan, the *nor'-westers* of the New Zealand Alps, and the *chinook* of North America.

The high plains at the foot of the Rocky Mountains, both in Canada and in the USA, owe their comparatively mild winters to the frequency of the *chinook* winds, which are less damaging than the *föhns* of Europe. They produce spells of weather with temperatures well above freezing, and although 40° F is seldom exceeded the warmth feels like midsummer by comparison with the previously much colder spells of weather. The *chinook*, which occurs when barometric pressure is high over the plateaux to the south and low over Canada, has important economic value in that it causes the snow—which is never very deep in these rather arid regions—to disappear as though by magic.

Not unlike the *föhn* and *chinook* in some respects are the *berg* winds of South Africa, which blow from the interior plateau of the continent towards the coasts. They are particularly characteristic of the winter when barometric pressure is high to the north and low to the south, and in the Cape region they occur roughly fifty times a year. Their frequency decreases somewhat as one travels towards the south-east coast, but their effects are felt along the whole of the plateau edge. Not only are these winds highly charged with dust but they are capable of producing temperatures exceeding 100° F, and, as a result, winter sometimes produces the highest temperatures of the year.

Those who live in southern California and in the Sacramento Valley of northern California experience identical winds known respectively as *santa annas* and *northers*. These bring clouds of dust from the high interior plateau, and the increase in temperature brought about by their descent causes a disastrous drying up of vegetation as well as acute discomfort to human beings and animals. Like the *berg* winds of South Africa, they occur chiefly in winter, but do their worst damage in spring when fruit trees are in bloom and when young fruit is formed.

Similar effects occur along parts of the Chilean coast. Near Concepcion, in latitude 37° S, the winds blowing down from the Sierras have given temperatures exceeding 100° F, and these are considerably higher than those of tropical Chile, where conditions for the development of *föhn* winds are less favourable.

During the winter, depressions entering the Mediterranean from the Straits of Gibraltar or from Biscay tend to draw in winds from deep in the Sahara. These are generally very hot and dry and heavily laden with penetrating red dust. They are given local names: the *scirocco* in Algeria, the *leveche* in Spain, and the *khamsin* in Egypt. They all produce acute bodily discomfort and mental strain, and vegetation withers. When the wind is descending, as it does near the Algerian coast or on the north coast of Sicily, the heat and aridity are further increased and maximum temperatures may then exceed 110° F. In crossing the Mediterranean the winds pick up moisture and are often damp and enervating by the time they reach the European coast, except for the south coast of Spain, where the sea passage is negligible.

At the rear of a depression moving west to east in the Mediterranean the wind may veer to the north-west or north, bringing in very cold air, and when this is channelled into narrow valleys its speed is increased and it becomes more squally. In France the wind is known as a *mistral*. It is particularly fierce in the Rhône valley, where speeds of over 85 mph have been reached, and on the coasts of Languedoc and Provence. Under similar conditions a bitterly cold north-east wind, known as the *bora*, is experienced on the eastern shore of the Adriatic, particularly in the north of this region, where it may last for several days. During the *bora* the wind speed at Trieste has averaged over 80 mph, with gusts exceeding 125 mph, temperatures have fallen to 14° F and the relative humidity to only 15 per cent.

In both hemispheres there are certain sub-tropical zones that experience a great range of temperatures. This is due to the movements of storm centres that draw in winds from latitudes far to the north or south, and the regions most affected by these winds lie on the eastern sides of continents.

One of the very unpleasant hot winds affecting these zones is the *brickfielder* of Victoria, Australia—a northerly flow that brings clouds of choking dust and temperatures up to 120° F. Prolonged hot spells, with temperatures of over 100° F are in fact an unpleasant feature of the climate of Melbourne. Another very distressing wind is the *zonda* of the Argentine, which, aided by exceptionally high humidity, brings a feeling of complete prostration.

Typical very cold winds of polar origin, that can invade the sub-tropics and lower the temperature by 30–40° F within 24 hours and are often accompanied by thunder and violent hail-storms and squalls, are the *southerly burster* of New South

Wales, the *pampero* of the Argentine, and the *norther* of the Gulf Atlantic States of the USA. Their arrival is generally very sudden and their power of destruction considerable.

In America cold winter winds from high latitudes may reach as far south as the Mexican coast, where they are called *nortes*, or, if they penetrate the high inland plateau, *papagayos*. The cold that they bring is intense, considering the latitude, and the fall of temperature tends to be very sudden and unpleasant. Cold winter winds known as *friagems* and *surazos* are even experienced in certain tropical regions of the South American continent; they can produce temperature falls of up to 50° F.

A wind that is particularly well known, due to its powerful climatic and economic effects in southern Asia, is the *monsoon*, a name derived from an Arabic word meaning 'season'. It was used originally to describe the winds of the Arabian Sea which blow for approximately six months from the south-west and six months from the north-east, but it has since been extended to cover certain other winds of marked seasonal persistence in many of the world's tropical regions. The primary cause of these winds is the great difference of temperature between land and sea areas during the respective summer and winter seasons; they are wet if they blow from sea to land and dry if they blow from land to sea.

The summer *monsoon* of India, Burma, and south China is caused by the heating of the Asian Continent, which causes air to rise on the grand scale in a cauldron-like action. To replace this, moist air from the ocean capable of producing sustained heavy rainfall is drawn in at surface level and travels across a vast oceanic fetch from well to the south of the equator.

Over the Mediterranean the prevailing winds between mid-May and mid-October are dry and blow from a northerly or north-easterly point towards the low-pressure areas of Arabia and India; they are strongest and most constant in the eastern Mediterranean, where they are known as *etesian* winds. The ancient Greeks were very familiar with them. Their average speed varies between 10–30 mph, but at times they reach 45 mph.

Etesian winds are normally stronger by day than by night, and when they blow over land they bring clouds of dust; as they cross the sea they raise stormy foam-topped waves that contrast strangely with the deep blue of the cloudless sky above. Sailing can be dangerous, particularly to windward of exposed rocky coasts. On the southern shores of the Mediterranean the northerly winds, having travelled some distance and picked up a considerable amount of moisture, bring mist and fog. This tends to be very persistent in Algeria and Tunisia and has a moderating effect on the temperatures in these regions.

Cool downslope winds that occur at night after it has been calm during the day are called *katabatics* and are particularly common in Iceland, Greenland, and Scandinavia during anticyclonic weather when barometric pressures are high and steady. The snow-covered uplands then cool rapidly, and the air in contact with them becomes cold and heavy and then gravitates to lower levels.

APPENDIX 3

NOTABLE WEATHER EVENTS IN BRITAIN

GALES

(Including certain floods known to be caused by gales.
Blizzards are listed separately.)

1 October 1250	Winchelsea was destroyed by a storm. Holland and Flanders also affected.
30 September 1555	Holinshed records that 'by occasion of such great wind and rain . . . the King's palace at Westminster and Westminster Hall were overflown with water'.
27 August–3 September 1588	A number of ships of the retreating Spanish Armada were wrecked after being driven off course by strong south-westerly winds as they sailed past Scotland and Ireland. Poor navigation was a contributory factor. Surviving ships encountered the strongest winds of the period between 2–3 September as they passed near to the Lizard. Part of English fleet in the Downs harassed by same storm (see p 173).
2–6 September 1666	The Great Fire of London that devastated 400 streets and lanes, 13,200 houses, St Paul's Cathedral, 89 parish churches, the Guildhall and other public buildings was fanned by a very strong easterly wind.
26 November 1703	The air was 'full of meteors and vaporous fires', said Daniel Defoe, who was an eye-witness of the great storm of this day. Thought to be the worst since early days of recorded history. Many English towns devastated, hundreds of mansions blown

down, hundreds of thousands of trees uprooted, fleets cast away at sea. Eddystone lighthouse destroyed and its designer, Winstanley, killed. Record high tides in Severn and Thames (see p 176).

1 November 1740 — London was struck by a hurricane which lasted from 6–11 pm and blew down one of the spires of Westminster Abbey.

10 November 1810 — Sea water driven against the North Sea coast by a persistent easterly gale held up the flow of rivers and flooded Boston, Lincs.

6–7 January 1839 — Many ships were wrecked and houses blown down in a gale known at the time as the 'Great Wind'. Menai suspension bridge damaged. Houses blown down in southern Scotland and northern England.

25 October 1859 — The *Royal Charter* was wrecked on the Anglesey coast in a severe gale, with the loss of nearly five hundred lives. Old chain pier at Brighton destroyed. Gale warnings based on telegraphed reports issued shortly afterwards for first time (see p 41).

21 February 1861 — One wing of the Crystal Palace and the tower of Chichester Cathedral were destroyed in a violent storm over southern England.

1–3 December 1863 — England and north-western France were struck by a violent gale.

24 March 1878 — The training ship HMS *Eurydice* foundered off the Isle of Wight, with the loss of all hands, in a sudden north-westerly gale.

28 December 1879 — A train carrying seventy-five persons was lost in the river when the then recently opened Tay Bridge was destroyed by a gale.

13–14 October 1881 — Buildings all over Britain were damaged in a very severe gale. Many ships lost at sea.

1–2 September 1883	Extensive damage occurred in western and southern England as a result of a violent gale that began its life on 24 August as a West Indian hurricane.
26 January 1884	A great storm produced the lowest sea-level barometric pressure ever recorded in Britain, 27·36 in (926·5 millibars) at Ochtertyre, Perthshire.
8–9 December 1886	Many ships lost in an intense storm that crossed northern Ireland and England from west to east. At Belfast barometer fell to 27·38 in (927·2 millibars).
24 March 1895	Fourteen people were killed and many hundreds injured in one of England's most destructive storms—the worst of the nineteenth century in the Midlands. 100 mph winds. Factory chimneys and church steeples blown down. Stationary train set in motion by wind led to a collision.
26–27 February 1903	A train was blown over Leven Viaduct, near Ulverston, Lancs., by 90 mph winds. Extensive structural damage to property.
5 December 1906	Scotland's first wireless tower, at Machrihanish, was blown down by a violent northwesterly gale.
2–3 December 1909	The Manx steamer *Ellen Vannin* sank during a gale, with much loss of life. Other ships wrecked as well. Barometer fell to 28·03 in (949 millibars) at Spurn Head.
28 October 1927	A curious gale led to a widespread breakdown of electrical power. Little rain, but 96 mph wind recorded at Southport and much sea spray carried inland leading to formation on insulators of salt crystals which liquified on following day.
6–7 January 1928	A north-westerly gale caused a great storm surge, leading to one of the worst Thames floods on record.

12 January 1930	A severe gale occurred which was one of the most destructive on record in England and the Channel.
9–10 November 1931	The Isle of Wight was for a time split in two by a gale-driven high tide between Yarmouth and Freshwater Bay. Much damage to Sussex coast bungalows at Shoreham and to Ship Inn, Winchelsea, by waves.
16–17 September 1935	Winds of 98 mph were recorded in Cornwall and 88 mph as far inland as Cardington, Beds, in a gale that was particularly severe for so early in the season. Windows in Cornwall coated with thick brown deposit consisting of mixture of salt from dried spray and scum from waste oil on the sea.
4 June 1944	A Channel gale made it necessary to postpone the Normandy invasion planned by the Allies in World War II. The fleet sailed two days later with improved but still rough weather.
17–22 June 1944	A second invasion gale wrecked landing craft on Normandy beach-head. Long strings of Mulberry tows lost, breakwaters smashed (see pp 185, 187).
30 December 1951	Violent gale-force winds occurred over the north of Scotland.
15 January 1952	This was almost a duplicate of the storm of sixteen days before, but it was even more violent. 100 mph winds in places caused extensive damage Orkney and Caithness, also in Shetlands and Hebrides.
17 December 1952	A squall wind of 111 mph was recorded at Cranwell, Lincs—at that time the highest officially recorded wind speed for an inland station. Severe gale over whole of Scotland and northern England.

31 January– *1 February 1953*	The *Princess Victoria* foundered in the North Channel on 31 January in a severe northerly gale. On 1 February great mass of North Sea water was driven southwards, causing disastrous flooding on the east coast of England and the coast of Holland, with much loss of life.
6 March 1967	In the Cairngorms a 145 mph gust of wind occurred during a severe gale. This is the highest officially confirmed wind speed ever recorded in Britain.
14–15 January 1968	Twenty people were killed in the Glasgow hurricane that occurred, unexpectedly, during the night. Damage to property in Clydeside area over £20 million. South-westerly wind accentuated in Clyde valley due to funnelling effect created by surrounding high ground.
17–20 March 1969	The Longhope (Orkney) lifeboat was lost, and its crew perished, in a severe easterly gale on 17 March. On 20 March Independent Television's 1,270 ft main transmitting mast for Yorkshire at Emley Moor collapsed under weight of accumulated ice resulting from gale and sub-zero temperatures.

NOTABLE FLOODS

27 January 1607	Disastrous flood in the Severn estuary overwhelmed many valleys. Many hundreds of people drowned; churches and many other buildings submerged.
8 May 1663	According to Pepys, 'extraordinary floods in a few hours, bearing away bridges, drowning horses, men and cattle'.
1 December 1768	The Thames and other rivers overflowed their banks. The Exeter coach, with six

	passengers and four horses was carried away by the flood near Staines.
24–26 January 1849	Great damage caused to Inverness by flooding. Loch Ness rose 14 feet, a height unequalled in its annals.
2 February 1852	The village of Holmsforth was almost destroyed, with great loss of life, when the Bilberry dam, near Huddersfield, burst after a period of very heavy rainfall.
mid-September 1852	At Putney the towing path was six feet under water and the Great Western line was flooded for four miles between Hanwell and Paddington. Hearse and horses upset in flooded Bath Road at Maidenhead on occasion of Duke of Wellington's funeral.
11–13 May 1886	A serious flood in the Severn valley caused by widespread and persistent rainfall.
28 January 1892	Severe floods in the Strathglass, Bonar Bridge and Strathspey areas. Bonar Bridge destroyed, parts of railway in Strathspey washed away.
12 July 1900	Ilkley, Yorks, devastated by a flood.
13–15 June 1903	Widespread floods caused by the longest period of continuous rainfall (58½ hours) ever recorded in Britain.
26 August 1912	Twenty-four hours of rain in East Anglia, producing over 8 in of rain. Floods reached their highest known level in Norwich.
28–29 June 1917	Bruton, Somerset flooded. 9·56 in recorded.
20 May 1920	A deluge fell on the Wolds west and south-west of Louth, Lincs. Torrent 200 yd wide swept through the town, drowning 22 people.
1–4 October 1920	Destructive floods in the mountains round Balmoral and Ballater, with over 7 in rainfall recorded.

8 July 1923	This was another famous Scottish deluge. Six hundred yards of railway track on high embankments disappeared into the torrent of water near Carrbridge.
31 May–1 June 1924	Severe floods in Shropshire and Worcestershire washed out the Worcester Agricultural Show.
28 June 1928	Two air currents meeting over the Welsh mountains produced 7·77 in of rainfall and extensive flooding at Blaenau Festiniog, Merionethshire.
12 August 1948	Heavy rainfall occurred in many parts of Britain. Flood damage of around £1 million in the Tweed valley.
15 August 1952	Lynmouth was flooded. 9·1 in rainfall recorded at nearby Simonsbath. In small basin of the Lyn rivers covering 39 square miles, flow of water nearly twice as great as the highest ever recorded in the Thames. Depression that caused flood moved from Britanny and at 6 pm on 15 August was centred over Exmoor.
10–11 July 1968	Seven people were killed in the West country floods. Five main bridges washed away. Worst hit areas south-east Devon and around Bristol and Bath. Floods also in Midlands. 7 in rainfall in 13 hours at Bath, 5·14 in in 24 hours at Gloucester and 2·4 in in 24 hours at Birmingham—its heaviest fall on record.
14–15 September 1968	Rain became torrential during the night of 14 September and continued throughout the following day. Rainfall totals ranged from 3 in in London area to 4 in in south Essex and 7½ in in Kent. Weald of Kent was transformed into vast lake. 1,000 houses flooded in Weybridge, Byfleet and Addlestone areas when river Wey burst its banks. Dormitory towns of Walton and Molesey

turned into lakes 5–6 ft deep when river Mole burst its banks. Widespread flooding in many other regions of eastern and southern England.

NOTABLE THUNDERSTORMS, HAILSTORMS AND TORNADOES

21 October 1638 A savage combination of thunderstorm and tornado swept through south-west England. Widespread damage. Tower and roof Widecombe-in-the-Moor church on southern flank of Dartmoor, demolished during divine service, killing many of congregation.

1 August 1846 Seven thousand panes of glass were broken in the Houses of Parliament in a series of violent hailstorms.

18 April 1850 A notable thunderstorm over Dublin was accompanied by a tornado that carried off roofs and blew down many chimney-stacks and trees. Very large hailstones.

2–3 August 1879 Hailstones up to six inches round fell in the Kingston and Ealing areas, smashing through roofs, in this historic London thunderstorm. Lasted throughout night, lightning flashes every second or less.

4 September 1886 Swansea was seriously damaged in a tornado. Kelvey Hill struck by waterspout, resulting in great currents of water rushing down slopes, bursting through houses. Forty families homeless.

9 June 1888 The village of Langtoft, Yorkshire, was partly destroyed by a tornado.

3 July 1892 A thunderstorm and cloudburst occurred nead Driffield, Yorks. Sheet of water 100 yd wide and 3 ft deep ran at 20–30 mph from higher ground, sounding like the roaring of the sea.

22 July 1907 Widespread heavy thunderstorms were particularly violent in South Wales where hailstones were nearly the size of hens' eggs.

9 June 1910 In a crop of violent thunderstorms between central Wales and Surrey hail fell so abundantly that it was piled up by the wind in heaps 2–3 ft high in places.

31 May 1911 Lightning occurred almost continuously for several hours at Epsom on Derby day. 17 people and 4 horses killed.

27 October 1913 A violent tornado of the American type crossed south Wales and passed on to Shropshire and Cheshire. Slates from roofs were found buried in trees. Severe thunderstorms between Exmouth and Chester.

16 June 1917 $4\frac{3}{4}$ in of rainfall was recorded at Campden Hill, Kensington in 2 hours in London's heaviest thunderstorm deluge. Yet no rain fell at London Bridge or Clapham Common.

16 July 1918 A hailstorm moved in a mile-wide straight line between Holmwood, Surrey and Bromley, Kent.

9–10 July 1923 6,924 flashes of lightning occurred in 6 hours in a London storm that raged throughout the night.

18 August 1924 A thunderstorm over Cannington, Somerset, produced 9·40 in of rain, most of which occurred within $4\frac{1}{2}$ hours.

20 June 1933 Severe flooding and damage by hail and lightning occurred during the Merseyside storm. Water over 10 ft deep at Bootle where a cyclist was swept away and drowned.

12–19 May 1935 Polar winds, with hailstorms, sleet and snow severely damaged flowers, fruit and vegetables in valleys of Herts and Bucks. Temperatures fell to 10° F at Rickmansworth and 13° F at South Farnborough on 17 May.

6 May 1936 A violent thunderstorm in Bedfordshire produced hailstones one inch in diameter. Torrential rain washed down much soil from hillsides, drowning many pigs.

27 June 1947 Pitch darkness in London at midday was followed by thunderstorms and squall winds of 64 mph.

18 July 1953 Burlington House in Piccadilly was struck by lightning. Part of main archway crashed to ground. Widespread thunderstorms elsewhere, with extensive damage and some loss of life.

18 July 1955 A cloudburst over Dorset and Somerset produced the highest 24-hour rainfall total ever recorded in Britain: 11 in at Martinstown, near Dorchester.

9–11 July 1959 A prolonged heat-wave was broken by severe hailstorms. Damage particularly great at Wokingham, Surrey.

21 July 1965 A tornado that formed near the river Wey swept through the gardens of the Royal Horticultural Society at Wisley, Surrey, damaging fruit plantations and uprooting many trees. Crossed A3 London to Portsmouth road and finished at Wisley airfield where hangers were damaged.

13 July 1967 A series of violent thunderstorms in western areas of Wiltshire and neighbouring parts of Somerset and Gloucestershire produced hailstones of over 2 in diameter.

21 April 1968 — A tornado travelled for some 3 miles from the Coventry outskirts to the village of Barnacle, destroying buildings, caravans and motor vehicles.

NOTABLE BLIZZARDS

February 1799 — A woman, buried for eight days in a snowdrift between Cambridge and Impington, was rescued alive.

January 1814 — A month of repeated heavy snowfalls.

29 October 1836 — Twelve in of snow fell during a remarkable blizzard for so early in the season. At Bury St Edmunds and in London and other regions it did not thaw for between 5 and 6 days.

19 April 1849 — The Westerham coach was buried in a snowdrift on Titsey Hill, Surrey, during a blizzard.

18–20 January 1881 — In this historic blizzard, rail and road traffic was paralysed over thousands of square miles in southern counties. Numerous families had to be dug out of their homes.

9–12 July 1888 — Widespread snowfalls occurred throughout Britain. These were heavy in upland districts in the north.

9–13 March 1891 — Many trains were buried in what was one of the fiercest blizzards of the century in southern England. The 'Zulu' express left Paddington for Plymouth at 3 pm on 9 March and arrived at its destination on 13 March, having spent four days in a large drift at Brent.

18 May 1891 — Holidaymakers were able to toboggan on

Whit Monday as a result of a heavy, drifting snowstorm over many hundreds of square miles in England.

26–30 December 1906	In a succession of snowstorms, which were particularly fierce in Scotland, Aberdeen was isolated for three days.
28 March 1916	A blizzard in southern England blew down trees and telegraph wires. Traffic was dislocated. Unusually large snow drifts for so late in season.
2 April 1917	Ireland experienced one of its fiercest and most destructive blizzards within memory. Ten-foot drifts on Galway coast within two hours after storm's onset.
25–26 December 1927	After more than a week of bitterly cold weather came the great Christmas snowstorm that is still remembered by many people. Snowdrifts halted traffic and isolated many villages.
27 February–1 March 1937	The term 'blizzard' was used officially for the first time. Thirteen-foot snowdrifts in Wales and northern England blocked roads, brought down telephone wires.
29 January 1940	Snowfall along the western flanks of the Pennines and the Lake District was so heavy that the southbound LMS express was held up for 36 hours a few miles south of Preston.
18–20 February 1941	A severe blizzard in the north-east gave 30 in of snow at Newcastle and 42 in at Durham.
29 January 1947	In a severe north-easterly blizzard the maximum temperature at Plymouth was 23° F and at the Scilly Isles only 28° F although the nearby seawater at the time had a surface temperature of nearly 50° F.

February 1947	Frequent catastrophic blizzards occurred throughout the month, as in January 1814.
25–26 April 1950	Six in of snow fell on the North Downs in a severe blizzard. Over 1,200 telephone poles and many trees brought down.
February 1953	After the prospect of a general thaw the whole of Britain was struck by a fierce blizzard when warm Atlantic air came into contact with bitterly cold air from the Continent. Many towns and villages isolated.
29 December 1962	Great drifts of snow occurred over south-west England and the English Channel coast.
1–4 January 1963	Hundreds of schools were forced to delay opening of new term by continuous blizzard conditions that moved northwards from the English Channel.
6 February 1963	Thirty-six hours of continuous heavy snow, 20-ft drifts, and winds of up to 103 mph were features of this blizzard—considered to be worst of the winter.
15 February 1963	Parts of Cumberland received heaviest snowfall within living memory.

NOTABLE FROSTS OF THE LAST 100 YEARS

December 1879	In this very cold month Greenwich recorded a mean temperature of 32·5° F and the record minimum screen temperature for the British Isles of −23° F on 4 December was recorded at Blackadder, Berwickshire.
January 1881	Edinburgh experienced 13 consecutive days with the air temperature continually below freezing point. Many trees killed. Rivers

frozen throughout country and large extent of Lough Neagh in Ulster frozen over.

5 January–19 March 1885–86
In this 60-day frost there was skating until 18 March.

December–January 1890–91
1890 produced the coldest December of the century. Thames nearly blocked by ice in London between Westminster and London Bridge. Heligoland cut off from mainland by ice, Baltic icebound, ice-blocked ships and harbours at Toulon, Bordeaux and Lisbon. River Tagus at Lisbon frozen over.

December 1894–February 1895
This was the finest skating winter of the century. Greenwich recorded a February mean temperature of 29·1° F. Skating on Thames. Special trains for skaters to Lake Windermere and Loch Lomond.

10 January–16 February 1916–17
Greenwich recorded a mean temperature of 31·6° F during this period. The winter was the most severe during World War I but it was milder in Scotland than in Ireland and the South and West of England.

February–mid-March 1928–29
During this intensely cold spell there was skating on Derwentwater which continued into March until the thaw. Temperatures then only a little below 70° F. Records state as many as fifty thousand people on the ice on Lake Windermere at one time.

December 1939–February 1940
Parts of Southern England enjoyed skating at Christmas, and this continued for six weeks until mid February. Lowest mean January temperature for 100 years at Greenwich of 30·8° F. Sea froze along south and south-east coasts, ice extending hundreds of yards offshore. Ice in Thames. Freezing rain on 27–28 January encased telephone wires in cylinders of ice of over an inch in diameter, causing great damage.

December 1940–January 1941	Sub-zero temperatures were frequent, and skating was widespread in Scotland. January the coldest month, as in previous winter.
January–February 1942	In this third successive cold winter both January and February were bitterly cold. February the coldest over England and Wales since 1895.
19–30 January 1945	Continuous frost was experienced at Whipsnade Zoo, and sub-zero temperatures were recorded in many parts of Britain. Beer froze in public houses and sea ice formed in Folkestone harbour. In London hammers of Big Ben became frozen and clock failed to chime properly. Icebreakers used on Grand Union Canal in Leicestershire. Excellent skating on Loch Lomond and in Lincolnshire.
December 1946–March 1947	The weather was snowy and extremely cold throughout the winter, but particularly in February, when the mean temperature at Kew was 30·0° F, nearly 11° F below average and only 0·9° F less cold than February 1895.
February 1956	On balance this was a colder month than February 1929 but somewhat less cold than February 1947. Snowfall was frequent, and was unusually heavy in south-west England. Minimum temperature 5° F Shawbury, Salop, night of 2 February, 3° F Wye College, Kent, night of 16 February.
27 December–27 February 1962–63	This winter produced the coldest January since 1795, and except for a break for a few days during the fourth week, day temperatures rose little above freezing. Most of country snow covered throughout month. Whole of February bitterly cold except for short intermission during second

week. Widespread and severe snowstorms throughout winter.

NOTABLE HEAT-WAVES

8 July 1707	It was so hot and airless that a number of men and horses died of heat-stroke at harvest work, and for some time afterwards the day was recalled as 'hot Tuesday'.
22 July 1858	100·5° F was registered at Tonbridge, Kent.
15 July 1881	101° F was recorded at Alton, Hants, and 100° F at Alderbury, near Salisbury, Wilts.
31 August–3 September 1906	Maximum daily temperatures at Greenwich were, respectively, 94·3° F, 91·9° F, 93·5° F and 91·0° F. In records covering nearly a century this proved to be only instance of four consecutive days giving temperatures in the nineties.
9 August 1911	The temperature reached 100° F at Greenwich in London's hottest day on record. Summer of 1911 was hottest in England since 1868 and was a factor that quickened the pulse of revolt in the fierce series of strikes and lock-outs of that year.
8 September 1911	Temperatures rose to 93° F in many districts of southern and central England and to 94° F at Greenwich and in a part of Northamptonshire.
11–13 July 1923	Temperatures exceeded 90° F on three consecutive days.
18–20 August 1932	Temperatures exceeded 90° F on three consecutive days, and on 19 August reached 98·9° F.
29 May–3 June 1947	Temperatures reached or exceeded 85° F on these six consecutive days.

16–18 August 1947	Temperatures of above 90° F were recorded in a number of places, and Bournemouth's maximum of 93° F on 16 August was the highest ever recorded there.
27 August 1947	80° F was recorded at Cape Wrath, on the north coast of Scotland, on this day.

NOTABLE FOGS

27 December–2 January 1813–14	London was covered by a dense fog. Impossible to see across streets, candles burnt in shops and counting-houses. Heavy coal tar pollution produced smarting effect on eyes.
7–13 December 1873	A thick fog occurred during London's 'Cattle Show' week. Forty per cent increase in death rate.
25–28 December 1944	In this famous Christmas fog visibility in London was virtually nil throughout the whole of Christmas Day and from midday on Boxing Day until 6 am on 27 December.
22 November–1 December 1948	Dense fog extending from the Thames estuary to the Welsh Borders and northwards to Lancashire and Yorkshire. Spread by 28 November in a 1,200-mile belt across north-west Europe from Wales to Finland. Nearly all shipping stopped. Airliners grounded in London, Paris, Copenhagen, Stockholm and Helsinki, and Berlin 'airlift' halted. Several train collisions. Bus services suspended for a time in the London area. When fog cleared on 1 December the *Queen Elizabeth*, *Queen Mary* and *Aquitania* sailed in convoy from Southampton.
5–10 December 1952	Four thousand people died during the great London 'smog'. Began as an ordinary water

fog, but was converted into smoke fog by pollution from chimneys. At County Hall weight of smoke in the air increased from 0·49 milligrams per cubic metre on 4 December to 4·46 on 7 and 8 December.

3–6 December 1962 This London fog was similar in many respects to that of December 1952. 750 people died.

APPENDIX 4

THE WORLD METEOROLOGICAL ORGANISATION AND THE WORLD WEATHER WATCH

The International Meteorological Organisation, which was formed in 1878, has made possible the voluntary interchange of vital weather data between nations, and, with the growth of civil aviation and the increasing importance of weather forecasts in national economies, its functions have expanded considerably over the years. The members of IMO used to be the directors of individual national meteorological services, but after the Second World War it was considered necessary to strengthen the organisation by making it an inter-governmental body. So in 1951 the World Meteorological Organisation was founded. It acts as an international exchange for national meteorological services, and its decisions are mandatory for all member countries. At its headquarters at Geneva it is supported by eight technical commissions, each responsible for making recommendations in its own branch of meteorology.

WMO stipulates the observations to be made for international exchange: the times of the observations, the elements to be observed, the units and codes to be used, and the telecommunications arrangements. This has achieved international standardisation, but there is still much work to be done. Meteorological services have not all reached the same state of development; a number of new or under-financed services are comparatively backward. There are still many gaps in the world network of observational stations, particularly over the vast areas of ocean away from the main shipping lanes.

The dual requirements of the meteorological forecasters and theoreticians led to the concept of the World Weather Watch, a revolutionary new scheme for expanding the world's weather observational system by using satellites, automatic weather stations, and a much greater number of conventional instruments. Provision has been made for a number of new ocean and island weather stations. Another function of the World Weather Watch has been to create a new system of world meteorological centres, which have been located at Melbourne, Moscow, and Washington. These are being supported by a number of new regional meteorological centres. Finally, the new scheme is promoting the better use of electronic computers for data-processing and is stepping up and improving telecommunications.

To ensure the success of the new scheme the WMO has launched a technical co-operation programme in which more than 100 experts are helping some sixty countries to build up their meteorological services to the required standards, and most of the funds for this programme come from various United Nations technical aid schemes. Some indication of the scope of the world's meteorological services today can be obtained from the total annual figure that is spent on them, which runs at around £2,500 m.

APPENDIX 5

FORECASTING AND CLIMATIC SERVICES: COSTS IN RELATION TO BENEFITS

The gross annual expenditure of the British Meteorological Office is divided as follows:

Basic weather service	£1,700,000
Royal Air Force	£3,000,000
Civil aviation and weather ships	£1,300,000
Other specialised services	£400,000
Research	£800,000
Total	£7,200,000

This amounts to only 2s 6d per head of population compared with 6s in Australia, 10s in Canada, and 21s in the USA. The annual expenditure by weather services of the Federal Government of the United States which, since 1961, have increased eighteen times as rapidly as those of the British Meteorological Office, amounts to £155,000,000, but, excluding the heavy cost of the weather satellite programme and other aspects of international research, the figure is £103,000,000.

COST-BENEFIT STUDIES FOR THE UK

The Meteorological Office has checked the cost of certain specialised services against their economic benefit. The value of the daily weather forecast on television, the radio, or in the Press, has been put at 2d per family. This may be too low if one considers that the forecasts often prevent much waste of time, contribute to the general health and efficiency of the population, and help in the planning of leisure activities, but even on this basis the value amounts to £30,000,000 in a year or eighteen times the cost of the basic service for the general public.

The economic value of the main specialised services are calculated as follows and each is given an appropriate benefit/cost ratio:

Services to Civil Aviation

These are given in the form of forecasts, information and advice

1. They increase safety, reducing the risk of loss or damage to aircraft, passengers, and equipment through adverse weather.
2. They save money by allowing aircraft to follow the most economical routes, and help to minimise delays due to unfavourable weather.
3. They help to maintain regular flight planning and maintenance schedules, and the continued up-dating of information on terminal weather conditions prevents unnecessary diversions and cancellations.
4. The climatic data supplied assists in the planning of new routes and in the design of new aircraft.

The number of civil aviation forecasts issued in a year is between 500,000 and 600,000, and the average cost of each is £2. Roughly a quarter of the forecasts issued are for BEA, making the total cost of the Corporation's meteorological service approximately £370,000. The benefits are calculated as follows. BEA's actual weather losses are approximately £1,500,000 per annum, and it is estimated that the cost of diversions, cancellations, extra fuel, and loss of custom due to airline irregularity would increase by 300 per cent if no forecasts were available. Further, the provision for the insurance of passengers and aircraft would need to be increased by 25 per cent. In all, the value of the weather forecasting service is conservatively estimated at £2,400,000, making a benefit/cost ratio of 6·5 to 1.

BOAC cancels only 0·5 per cent of its flights due to bad weather, and it is estimated that the absence of weather information would increase the percentage to 1·5, which would be equivalent to an additional penalty—based on revenue loss and extra flying costs—of £50 per flight. The cost of diversions, extra fuel, and loss of payload would amount to another £25 per flight and the increase in insurance premiums to £35 per flight. The benefit of the existing weather service is in the efficient planning and monitoring of routes, leading to a saving of 3 per cent in flight time or 12½ minutes on every 7 hour flight, which amounts to an average saving of £125 per flight. The total benefit of weather services therefore amounts to an average of £235 per flight, which represents a benefit/cost ratio of 8 to 1.

The total annual economic benefit of meteorology to British civil airlines is estimated at £6,500,000, with an overall benefit/cost ratio of 10 to 1.

Services to Agriculture

The annual value of agricultural, horiticultural, and forestry produce in the UK is estimated at about £2,000,000,000. If all forecasts and meteorological advice were accurate and fully utilised, production would be expected to be at least 5 per cent higher—an increase that would be worth £100,000,000 per annum. At present the value of weather forecasting and climatic services are estimated to be worth 1 per cent of the annual value of gross production, or £20,000,000 per annum. The cost of the annual services amounts to only £50,000, which represents the very significant benefit/cost ratio of 400 to 1.

Services to Building and Civil Engineering

In the USA, where greater protective measures are taken against adverse weather than in Britain, time lost is estimated to be 2 per cent of production. In the UK it is 3.5 per cent of production, amounting to £100,000,000 per annum. If weather forecasts are used to plan interior work during bad weather and avoid its worst effects, a saving of at least one working day per man per year is feasible, amounting to £10 million per year for the whole industry. Based against the total of the varying charges made for specialised services to individual contractors, an overall benefit/cost ratio is envisaged of at least 40 to 1.

Services to the Electricity Industry

Without meteorological advice it would be necessary for the Central Electricity Generating Board to increase its reserve capacity by at least 10 per cent per annum, at a cost of £200,000. The forecasts received enable the CEGB to operate within the present reserve limits as well as to plan maintenance schedules and bring forward reserve equipment as necessary to meet demands that are immediately anticipated. The annual charge for the service of £12,000 represents a benefit/cost ratio of at least 20 to 1.

Services to the Gas Industry

Climatological and current forecast data are essential for predicting peak demand and in coping with fluctuations, without creating excessive storage capacity or risking breakdown of supplies. In one year a single Gas Board reckoned on reducing its storage capacity by at least 20 per cent, as a result of receiving meteorological information, which is equivalent to an annual saving of £50,000. For the UK as a whole, allowing for a relatively smaller annual average unit of saving for each Board due to the fact that some years are milder than others, the total saving for the industry has been estimated at around £300,000 per annum. Against a charge of £3,000 per annum, the benefit/cost ratio is 100 to 1.

General Climatological Services

Climatological data is used in a number of important projects, including the design of reservoirs and of urban storm-water drainage systems. The annual cost of the latter is £30,000,000 and the cost of providing the information—which is valid for 20 years—is £30,000. It is estimated that the information saves 1 per cent of the capital expenditure, amounting to a benefit/cost ratio over each 20-year period of 200 to 1.

APPENDIX 6

THE AUTOMOBILE ASSOCIATION'S ROAD WEATHER SERVICE

For administrative purposes the country is split up into thirty-five areas, each controlled by an AA office. Patrols in each area send reports on prevailing conditions by radio or telephone at stated times to their area offices, which in turn pass the information on to the Association's head office in London. Here a countrywide picture of road conditions is built up and the information recorded on a large-scale map. A special code of symbols is employed to show the various types of weather which motorists are liable to encounter.

Area offices also interchange information so that they are aware of conditions in neighbouring areas. When all area reports have been received in London and the countrywide picture has been built up, it is in turn transmitted to all area offices, as well as to the Meteorological Office, so that each one soon knows what conditions are like in the rest of the country.

Weather does not remain static for very long, and AA offices keep each other constantly informed of changes by sending amendment messages, but at least three times a day each office reports afresh on the condition of every road for which it is responsible.

A teleprinter network, linking offices up and down the country enables messages to be sent simultaneously to all areas or, if required, to selected offices only. A report sent in by an AA patrol can be collated with other reports and distributed to all offices in a matter of minutes.

During periods of exceptionally bad weather over 10,000

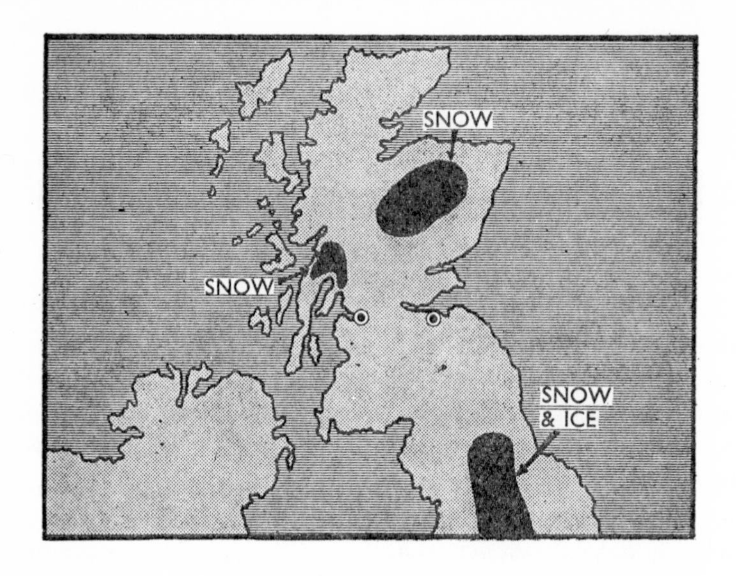

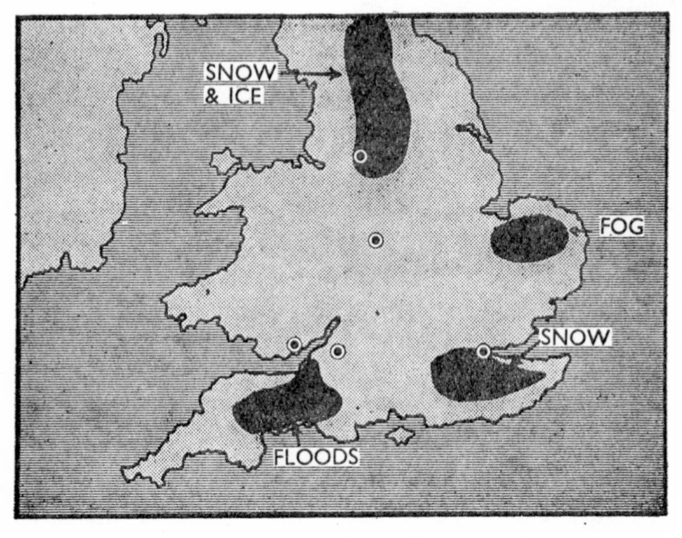

Fig. 22. Specimen AA charts of road conditions.

calls a day are handled by the London headquarters alone, and in an average winter the headquarters and area offices between them receive close on 1,000,000 enquiries.

Since 1958 motorists in thirteen areas have been able to dial a special number during the winter and listen to a recording giving an up-to-date report on the state of the road within a 50-mile radius. Bulletins prepared at the local AA offices are recorded on tape by the local telephone exchange and new recordings are made as changing conditions require.

AA weather information, though intended primarily for members, is also supplied as a national service to the Press and to all radio and television networks at regular intervals. The Association provides its own special charts for screening on television, which are shown during the evening for the benefit of travellers the following day.

The Met Office deals with forecasting of future weather, whereas the AA service is concerned solely with reporting existing conditions. For motorists, therefore, the two services are complementary.

APPENDIX 7

HOW TO FORECAST A FROST

A 'ground frost', according to the Meteorological Office, occurs when the thermometer on the grass has fallen to 30° F (−1° C) or below. But because damage may be done to plants before the temperature reaches this level and before there is any visible ice deposit on the ground, gardeners need to be particularly wary during the early growing months.

Frost warnings are issued by the official forecaster for broad general regions only. One can, however, estimate the risk of local ground frost in spring when the air has a polar or continental origin, when wind strength is below 5 mph, and when the sky is clear or nearly clear of cloud.

Make a note of the temperature at 3.30 pm and again at 10.15 pm, multiply the difference between the readings by two, and then subtract this figure from that of the 3.30 pm reading. The result will be the likely temperature at 5 am on the following morning.

The garden's situation will have a bearing on the severity and number of frosts during any one season. The higher any location, the less likely will it experience radiation frost, for cold air, being comparatively heavy, gravitates to the lowest levels it can find. The lowest night temperatures on frosty nights occur therefore in inland valleys, and some may be known as 'frost hollows' if the accumulating cold air forms into a deep pool.

If one compares present day records, with those of 40–50 years ago, springs do not appear to be colder now. But a patch of warm springs did occur in the period 1940–50, when the month of March gave mean temperatures of above 45° F (7° C) on four occasions, compared to only three occasions during the subsequent 16-year period from 1951–66. During 1940–50 five Aprils gave mean temperatures of above 50° F (10° C), but only April 1961 has beaten this level since 1950. May temperatures have not varied as much as those of the early spring, and mean temperatures since 1951 have not been below 50° F (10° C).

In most inland regions of Britain frost frequently shows little variation between the months of November and April, but there are a number of widely scattered districts—including south-east London, Glasgow, Oxford, Falmouth, and Liverpool which have a maximum frequency in March.

APPENDIX 8

MEANINGS OF WORDS USED IN WEATHER FORECASTS WHEN NO PRECIPITATION IS EXPECTED, OR TO DESCRIBE THE CHARACTER OF THE GENERAL WEATHER BETWEEN SHOWERS

For daylight periods

Fine	No precipitation or thick fog. Some sunshine.
Dry	No precipitation or thick fog.
Sunny	Much, or completely blue sky.
Sunny periods **Sunny intervals** }	Variable skies with considerable blue patches and corresponding sunshine.
Bright	A good deal of cloud, but considerable diffused sunlight or even occasionally some direct sunshine.
Bright periods **Bright intervals** }	Cloud rather variable in amount and thickness, but considerable diffused sunlight at times or even occasional direct sunshine.
Cloudy	Cloud nearly or completely covering the sky and mostly thick or dense enough to reduce the daylight appreciably below that described as bright.
Dull	Cloud completely covering the sky and thick or dense enough to reduce the daylight substantially below that described as bright.

At night

Fine **Dry** }	No precipitation or thick fog.

Clear	No fog, and little or no cloud.
Cloudy	Cloud nearly or completely covering the sky.
Variable cloudiness	Cloud varying between about one-quarter and three-quarters of sky covered.

APPENDIX 9

PRECAUTIONS TO TAKE IN A THUNDERSTORM

The chances of being struck by lightning are remote, but if you are one of those who feel happier to take precautions during a thunderstorm the general principles to observe are these. It is safer indoors than out; that is, inside any house or solid building. Wooden and other outbuildings offer little or no protection if struck by lightning. Not all parts of a house are equally safe. If a chimney is struck by lightning and no conductor is fitted, the stroke will be conveyed down the chimney breast, so that anywhere near the hearth could be an ill-chosen spot. Copper pipes attached to the plumbing installation may become live in the event of a direct stroke of lightning, so it is wise to keep away from the bathroom and the kitchen sink. For those who are very frightened during thunderstorms, the safest place in a house is normally under the stairs or in a ground floor room where there are no exterior walls. There is no extra safety to be gained by drawing blinds, covering up mirrors or removing metal-framed spectacles.

Outdoors a person becomes somewhat more vulnerable if he carries a long metal-tipped object such as an extended umbrella or a golf club. Lightning tends to strike the first substantial upright object in its path, so that open, flat areas such as golf courses, cricket pitches and race courses are in a bad position. Thunderstorms often travel up river valleys, so the banks of rivers should be avoided if possible, also exposed beaches; one should avoid sheltering under individual large trees or near metal fences.

When in a car it is best to make for the lowest ground in

the area, but it is comparatively safe inside a motor vehicle of all metal construction. If it is struck by lightning the current will travel from the metal framework to the wheel hubs and then discharge itself by jumping the gap between the hub and the ground, leaving the occupants inside unhurt.

Lightning conductors are advisable on isolated, large, or very exposed buildings, but one conductor will give limited protection only. Houses with several chimneys or a broad roof expanse may need several conductors, bonded together by copper strip. Each installation must be given a proper conductivity test, and then should be tested annually to ensure that it is still efficiently earthed.

Television aerials fitted with automatic earthing devices are not to be regarded as lightning conductors, being designed merely to carry to earth the minor atmospheric charges that build up during thundery weather. A direct stroke would melt the aerial, damage the roof, and possibly carry the charge inside the house.

APPENDIX 10

CONVERSION TABLES

FAHRENHEIT TO CENTIGRADE

	Centigrade									
Fahr	0	1	2	3	4	5	6	7	8	9
10	–12·22	–11·67	–11·11	–10·56	–10·00	–9·44	–8·89	–8·33	–7·78	–7·22
20	–6·67	–6·11	–5·56	–5·00	–4·44	–3·89	–3·33	–2·78	–2·22	–1·67
30	–1·11	–0·56	0·00	0·56	1·11	1·67	2·22	2·78	3·33	3·89
40	4·44	5·00	5·56	6·11	6·67	7·22	7·78	8·33	8·89	9·44
50	10·00	10·56	11·11	11·67	12·22	12·78	13·33	13·89	14·44	15·00
60	15·56	16·11	16·67	17·22	17·78	18·33	18·89	19·44	20·00	20·56
70	21·11	21·67	22·22	22·78	23·33	23·89	24·44	25·00	25·56	26·11
80	26·67	27·22	27·78	28·33	28·89	29·44	30·00	30·56	31·11	31·67
90	32·22	32·78	33·33	33·89	34·44	35·00	35·56	36·11	36·67	37·22

CENTIGRADE TO FAHRENHEIT

Fahrenheit										
–1	–2	–3	–4	–5	–6	–7	–8	–9	0	Cent
12·2	10·4	8·6	6·8	5·0	3·2	1·4	–0·4	–2·2	–4·0	–20
30·2	28·4	26·6	24·8	23·0	21·2	19·4	17·6	15·8	14·0	–10

Cent	0	1	2	3	4	5	6	7	8	9
0	32·0	33·8	35·6	37·4	39·2	41·0	42·8	44·6	46·4	48·2
10	50·0	51·8	53·6	55·4	57·2	59·0	60·8	62·6	64·4	66·2
20	68·0	69·8	71·6	73·4	75·2	77·0	78·8	80·6	82·4	84·2
30	86·0	87·8	89·6	91·4	93·2	95·0	96·8	98·6	100·4	102·2

INCHES AND MILLIMETRES

Inches	*Milli-metres*	*Inches*	*Milli-metres*	*Inches*	*Milli-metres*	*Inches*	*Milli-metres*	*Inches*	*Milli-metres*
0·05	1·3	3·3	83·8	6·6	167·6	9·9	251·5	30·3	769·6
0·1	2·5	3·4	86·4	6·7	170·2	**10·0**	254·0	30·4	772·2
0·2	5·1	3·5	88·9	6·8	172·7	11·0	279·4	30·5	774·7
0·3	7·6	3·6	91·4	6·9	175·3	12·0	304·8	31·0	787·4
0·4	10·2	3·7	94·0	**7·0**	177·8	13·0	330·2	32·0	812·8
0·5	12·7	3·8	96·5	7·1	180·3	14·0	355·6	33·0	838·2
0·6	15·2	3·9	99·1	7·2	182·9	15·0	381·0	34·0	863·6
0·7	17·8	**4·0**	101·6	7·3	185·4	16·0	406·4	35·0	889·0
0·8	20·3	4·1	104·1	7·4	188·0	17·0	431·8	36·0	914·4
0·9	22·9	4·2	106·7	7·5	190·5	18·0	457·2	37·0	939·8
1·0	25·4	4·3	109·2	7·6	193·0	19·0	482·6	38·0	965·2
1·1	27·9	4·4	111·8	7·7	195·6	**20·0**	508·0	39·0	990·6
1·2	30·5	4·5	114·3	7·8	198·1	21·0	533·4	**40·0**	1016·0
1·3	33·0	4·6	116·8	7·9	200·7	22·0	558·8	41·0	1041·4
1·4	35·6	4·7	119·4	**8·0**	203·2	23·0	584·2	42·0	1066·8
1·5	38·1	4·8	121·9	8·1	205·7	24·0	609·6	43·0	1092·2
1·6	40·6	4·9	124·5	8·2	208·3	25·0	635·0	44·0	1117·6
1·7	43·2	**5·0**	127·0	8·3	210·8	26·0	660·4	**45·0**	1143·0
1·8	45·7	5·1	129·5	8·4	213·4	27·0	685·8	46·0	1168·4
1·9	48·3	5·2	132·1	8·5	215·9	28·0	711·2	47·0	1193·8
2·0	50·8	5·3	134·6	8·6	218·4	29·0	736·6	48·0	1219·2
2·1	53·3	5·4	137·2	8·7	221·0	29·1	739·1	49·0	1244·6
2·2	55·9	5·5	139·7	8·8	223·5	29·2	741·7	**50·0**	1270·0
2·3	58·4	5·6	142·2	8·9	226·1	29·3	744·2	51·0	1295·4
2·4	61·0	5·7	144·8	**9·0**	228·6	29·4	746·8	52·0	1320·8
2·5	63·5	5·8	147·3	9·1	231·1	29·5	749·3	53·0	1346·2
2·6	66·0	5·9	149·9	9·2	233·7	29·6	751·8	54·0	1371·6
2·7	68·6	**6·0**	152·4	9·3	236·2	29·7	754·4	55·0	1397·0
2·8	71·1	6·1	154·9	9·4	238·8	29·8	756·9	56·0	1422·4
2·9	73·7	6·2	157·5	9·5	241·3	29·9	759·5	57·0	1447·8
3·0	76·2	6·3	160·0	9·6	243·8	**30·0**	762·0	58·0	1473·2
3·1	78·7	6·4	162·6	9·7	246·4	30·1	764·5	59·0	1498·6
3·2	81·3	6·5	165·1	9·8	248·9	30·2	767·1	**60·0**	1524·0

INCHES AND MILLIBARS

Equivalents in Millibars of Inches of Mercury at 32° F, Latitude 45°

Mercury Inches	*0·00*	*0·05*
	Millibars	
28·0	948·2	949·9
28·1	951·6	953·2
28·2	954·9	956·6
28·3	958·3	960·0
28·4	961·7	963·4
28·5	965·1	966·8
28·6	968·5	970·2
28·7	971·9	973·6
28·8	975·3	977·0
28·9	978·6	980·3
29·0	982·0	983·7
29·1	985·4	987·1
29·2	988·8	990·5
29·3	992·2	993·9
29·4	995·6	997·3
29·5	999·0	1000·7
29·6	1002·4	1004·0
29·7	1005·7	1007·4
29·8	1009·1	1010·8
29·9	1012·5	1014·2
30·0	1015·9	1017·6
30·1	1019·3	1021·0
30·2	1022·7	1024·4
30·3	1026·1	1027·7
30·4	1029·4	1031·1
30·5	1032·8	1034·5
30·6	1036·2	1037·9
30·7	1039·6	1041·3
30·8	1043·0	1044·7
30·9	1046·4	1048·1

APPENDIX 11

COUNTY RAINFALL ORGANISATIONS IN BRITAIN

A number of county rainfall organisations or associations have been formed in Britain. Many of the respective members are amateur meteorologists who contribute to the British Rainfall survey (see p. 226) which is a large undertaking and does not publish its annual charts until three years after each particular year under review. The members of the county associations aim to publish their records at frequent intervals and to submit these to local newspapers, river boards, contractors, consulting engineers, and to any others who are interested.

The addresses of the respective honorary secretaries of known county rainfall organisations are as follows:

D. W. Bogle, Cornwall Rainfall Association, Roughtor View, Camelford, Cornwall.

Squadron Leader T. B. Norgate, Norfolk Rainfall Organisation, Deighton Hills, Taverham, Norwich, NOR 53X.

In addition, the work of the former Devon Rainfall Organisation has been taken over by the Devon River Board. This distributes to all its observers a summary of each year's readings and exceptional observations. Particulars obtainable from Miss J. Ilett, Devon River Authority, County Hall, Exeter, Devon.

APPENDIX 12

MANUFACTURERS OF GENERAL METEOROLOGICAL INSTRUMENTS

Artech Meteorological Ltd, The Parade, Frimley, Camberley, Surrey.

Casella, C. F. & Co Ltd,* Regent House, Britannia Walk, London, N1.

Darton, F. & Co Ltd, Mercury House, Vale Road, Bushey, Herts.

Lufft Instrument Co Ltd, (London & UK agents: The Century Optical Co Ltd, 164–66 Tottenham Court Road, London, W1.)

Negretti & Zambra Ltd, 122 Regent Street, London, W1.

Surplice & Tozer Engineering Co Ltd, Acre Works, Windsor, Berks.

Walker, Thos & Son Ltd, 58 Oxford Street, Birmingham 5.

AGENT, AND CONSULTANT INSTRUMENT SERVICE

Meteorological Instrument Co, Wendover, Bucks.

* Also manufacturers of meteorological instruments for schools.

INDEX

Page references to illustrations are printed in italic